T0135860

Beyond Physics Content Knowledge

Modeling Competence Regarding Nature of Scientific Inquiry and Nature of Scientific Knowledge

Dissertation

von Irene Neumann, geb. Zilker

aus Würzburg

eingereicht zur Erlangung

des Doktorgrades der Naturphilosophie (Dr. phil. nat.)

an der Fakultät für Physik

der Universität Duisburg-Essen

1. Gutachter: Prof. Dr. Hans E. Fischer

2. Gutachter: Prof. Dr. Alexander Kauertz

3. Gutachter: Prof. Dr. Dietmar Höttecke

Tag der mündlichen Prüfung: 15. April 2011

Bibliografische Information der Deutschen Nationalbibliothek

Die Deutsche Nationalbibliothek verzeichnet diese Publikation in der
Deutschen Nationalbibliografie; detaillierte bibliografische Daten sind
im Internet über http://dnb.d-nb.de abrufbar.

ISBN 978-3-8325-2880-5

Logos Verlag Berlin GmbH
Comeniushof, Gubener Str. 47,
10243 Berlin
Tel.: +49 (0)30 42 85 10 90
Fax: +49 (0)30 42 85 10 92
INTERNET: http://www.logos-verlag.de

This doctoral project was conducted in scope of the Graduate School *Teaching and Learning of Science nwu-essen*. Financial support was provided by the German Research Foundation DFG.

Acknowledgements

Many people have supported me during the time I worked on this thesis. I thank each one of them. To name them all would not fit into this book.

I am very thankful to my supervisor Hans E. Fischer. You gave me the opportunity to conduct this Ph.D. project, and I deeply enjoyed working as your graduate student. You constantly challenged me, but never let me feel undefended; you guided me, but left me enough space to develop my own scientific thoughts and personality.

I am also very thankful to Elke Sumfleth who takes an important role in shaping the nwu-essen the way it is. I will always remember it as a place providing perfect conditions to write this thesis.

Doing this thesis allowed me to work with two wonderful people: Judy and Norm Lederman. I was honored to be allowed to collaborate with you; I thank you for your support, your continuous willingness to explain the concept of nature of science to me and the fruitful discussions we had; I am deeply grateful that you always made me feel as if I was one very special Ph.D. student of yours. I also thank you for your outstanding hospitality. I always felt welcome at your house. Thank you for perfect memories of my visits to Oak Park and Chicago – Go Bears!

I am also very thankful to 'my' post-doc Alexander Kauertz, who supported me – the time he was at Essen but also when he left to take the next step in his career. In particular, I am deeply grateful for the time you spent with me to discuss my thesis, the guidance you provided in difficult moments and finally, your on-a-short-notice proof reading.

One more wonderful person part of my U.S. memories is my overseas officemate, Gary Holliday. A big thank you, for all the help you provided to me – working with me on improving my test items, your thorough revisions on proposals and conference proceedings and your general support in any questions or problems I had or created; including your valuable fashion tips concerning Urlacher- and Cutler-Shirts.

The nwu-essen would not have been the perfect place it was to work at without all my colleagues. In particular, I thank you, Anna, for discussing many philosophical issues. Melanie, thank you for the time together and some Franconian flair in Essen. Sabine, I enjoyed singing and laughing with you. Thank you, Hendrik, for the many coffee breaks filled up with discussions. Annett and Jill, you provided me with statistical support and, much more importantly, lent a sympathetic ear in any emergency case. Finally, thanks to the pub-quizzers for an opportunity to think about something else than my Ph.D. project.

When I started at Essen, I was assigned an officemate; when I left, I left a friend. Markus, I will never forget the days and late nights we spent together at the office working on whatever. I am deeply grateful for your being a human dictionary that was always open for me. Thank you for always letting me bother you with work and non-work problems.

I am thankful to all students participating in my studies and to each teacher and principal who facilitated this participation. Thanks to the student workers who supported me during data collection and processing. Special thanks to Siv Ling Ley for your comprehensive search of physics history stories.

Thank you, Jon Higgins, for your thorough proofreading and your help on English language issues.

I am deeply grateful to my family – my parents, Julia, Veronika and Tante Anna. Away from home, I was not able to keep contact as close as I wished to do. Nevertheless, you always were there for me to hold me when I struggled.

Last but not least, I thank my husband Knut: Without you, I would never have started this project. I thank you for sharing the happy moments during this process of work; even more, thank you for encouraging me every time I needed it. Thank you for the enriching discussions that stressed your nerves and took some of what little spare time you had. I always felt that you had my back, even when you were far away.

Table of Contents

Introduction

In today's world, societies face serious decisions: Should we continue to utilize nuclear fission energy and how should radioactive waste be disposed; should we do something about global warming and if so what? These are just two of many issues that need to be addressed and that will impact the future of all citizens. Consequently, such issues are highly debated in public and private circles. Sometimes, findings of scientific studies are used in such discussions to support claims concerning a particular topic, such as whether salt domes are appropriate sites for radioactive waste disposal. Intriguingly, sometimes contradicting claims are inferred from the same study, or studies suggesting opposed conclusions are cited. Certainly, a thorough and deep understanding of science and technology will help in resolving such contradictions. However, not everybody is a scientist or engineer; neither should everybody have to be a scientist or engineer. For the average citizen, it is much more important to be able to make use of information provided by scientific studies. Such information may be applied in decisions of general interest, as for example whether the operating time of nuclear power plants should be extended, but also for individual decisisons, such as whether to use nuclear energy or regenerative energy in one's household. A requirement for using scientific information is the ability to judge scientific findings with regard to their scientific relevance and soundness: To what extent are the findings reliable and meaningful; how can it be that different studies suggest different conclusions and which study should be trusted? To do so, the average citizen needs to comprehend how scientific knowledge is generated and characterized. In science education research, an understanding of how scientific knowledge is generated and which characteristics it carries refers to an adequate 'understanding about the nature of science'.

Besides the issues illustrated above, there are many other examples for how citizens need to be informed about the nature of science for their own benefit but also for the benefit of society as a whole (cf. Hößle, Höttecke, & Kircher, 2004). Therefore, achieving an adequate understanding about the nature of science is considered a central aim of science education (e.g., Driver, Leach, Millar, & Scott, 1996; Kircher & Dittmer, 2004). Accordingly, several countries explicitly include the content of nature of science in their educational documents (McComas & Olson, 1998). However, the recently introduced German national science education standards do not (cf. Sekretariat der Ständigen Konferenz der Kultusminister der Länder der Bundesrepublik Deutschland [KMK, Secretary of the Standing Conference of the Ministers of Education and Cultural Affairs of the Länder in the Federal Republic of Germany], 2005a, 2005b, 2005c). This is remarkable since the German science education standards demand that students develop scientific competences to be able to responsibly participate in and contribute to society's science and technology-related discussions. In the standards' framework, competences represent more than mere knowledge; in fact, competences refer to the ability of applying particular knowledge and skills to solve particular problems. Abilities and skills that are needed in solving such problems as well as respective proficiency levels are summarized and detailed by competence models (cf. Klieme, Avenarius, Blum, Döbrich, Gruber, Prenzel, et al., 2004).

To sum up, (1) an understanding about the nature of science is important to the benefit of the individual as well as to the benefit of society as a whole, and (2) the German standards set a framework for providing students with competences to be responsible citizens, yet without explicitly including the concept of nature of science. The present project faces this desideratum and aims at providing a foundation upon which the standards may be expanded to include the concept of nature of science. In doing so, findings from research on competence modeling and on the nature of science are applied. Based on these findings, a competence model regarding nature of science, in particular regarding Nature of Scientific Inquiry (NOSI) and Nature of Scientific Knowledge (NOS) within the domain of physics, is developed and validated.

The first part details the two major research areas upon which the developed competence model builds: research on competence modeling and research on the nature of science. Moreover, how these research areas can be brought together is illustrated by developing a competence model regarding NOSI and NOS. Based on this theoretical background, research questions and hypotheses are derived in the second part. The third part details the operationalization of the theoretical competence model into test items, and describes the design of two studies to empirically validate the developed model. The fourth part presents the results of the studies, which are then discussed in the fifth and final part.

Theoretical Background

At the heart of this dissertation is a competence model regarding the Nature of Scientific Inquiry (NOSI) and Nature of Scientific Knowledge (NOS). Along with the introduction of national education standards, the development of competence models (i.e. competence modeling) became of particular importance in Germany and came into focus of a recently growing research area. The introduction of standards has to be viewed against the background of the German notion of education, *Bildung*. Within this traditional notion, science played a secondary role only, and thus was not viewed as being central to education (Fischer, Kauertz, & Neumann, 2008). However, as a result of German students' mediocre performance in studies of international school achievement comparisons, like the Third International Mathematics and Science Study (TIMSS) and the Programme for International Student Assessment (PISA), a change occurred in the German education system (cf. Helmke, 2000). This change has led to scientific literacy gaining importance in German education (Neumann, Fischer, & Kauertz, 2010); it has also resulted in the introduction of the German science education standards demanding particular competences as the outcome of compulsory secondary education (Fischer et al., 2008; Klieme et al., 2004). Parallel to this change in the German education system, new notions of *Bildung* were developed. They focus on educating students as future citizens to responsibly participate in and to advance societies including science and technology.

Due to the increasing relevance of science to modern societies and daily life, in Chapter 1, understanding the concept of nature of science is shown to be a reasonable and important component of contemporary notions of *Bildung*. The concept of competence is theoretically outlined in Chapter 2, which also details how competence models build the framework to define what competence means in a particular domain. In Chapter 3, Nature of Scientific Inquiry and Nature of Scientific Knowledge are demonstrated to be central educational contents. A review of current research illustrates that students hold inadequate conceptions about the nature of science and that competence modeling concerning the nature of science has not yet been approached. Consequently, a competence model regarding Nature of Scientific Inquiry and Nature of Scientific Knowledge is devised in Chapter 4.

1   Bildung *and Scientific Literacy*

For a long time, people have been trying to answer what the overarching goal of education is (cf. Reble, 2004). For the German speaking world, the quest to define the overarching goal of education is mainly connected with the idea of *Bildung*. The term '*Bildung*' is often translated as 'education', although both terms carry different meanings. *Bildung* corresponds to a particular idea or notion of education, for which different theories exist.

Classical Bildung

The notion of classical *Bildung*, which developed between 1770 and 1830, had a strong influence on German educational philosophy (Biesta, 2002; Klafki, 2007). Klafki (2007) identified four common characteristics of classical theories of *Bildung*. First, *Bildung* carries the meaning of self-formation. This aspect relates to the development of the individual personality of a human being, in particular to the personality's development related to humanitarianism. Second, *Bildung* is concerned with societal issues: One particular aim of *Bildung* is to promote society instead of encouraging with individual interests and concerns. Third, *Bildung* carries the idea of an interplay between individuality and community (*Gemeinschaftlichkeit*), in which individuality does not refer to isolation from other people but, rather, is obtained by interacting with them. Finally, *Bildung* is supposed to be characteristic of the broadly educated in contrast to the specifically trained. In summary, *Bildung* refers "to the cultivation of the inner life, that is, of the human soul, the human mind and the human person; or, to be more precise, to the person's humanity" (Biesta, 2002, p. 378). In this notion, *Bildung* was limited to a particular, small group of people: Mainly elite groups (e.g., reigning men, and future scholars) had access to *Bildung* while the broader public did not (cf. Tenorth, 1994).

This idea of classical *Bildung* has strongly affected the German education system (Fischer, et al., 2008; Kircher, 2007). The focus on the development of personality, humanitarianism and individuality led to an emphasis on the teaching and learning of classical languages and cultures (i.e., Latin and ancient Greek) and resulted in a neglect of the sciences (cf. Kircher, 2007). Science was viewed as dealing with realities that were not necessary to form an individual's human character. Accordingly, science education was virtually relegated to the so-called *Realgymnasium* which served to educate the middle class according to its typical professional needs; that is, *Realgymnasium* served to prepare the middle class for their mostly technical trade. For a long time, certificates earned at the *Realgymnasium* were not sufficient qualifications for university admission – in stark contrast to those certificates from regular *Gymnasium* (Kircher, 2007). This illustrates the past disregard of science in higher education. Accordingly, understanding the nature of science was no particular goal of higher education as well.

Scientific literacy

In the United States, science has carried higher relevance in the scope of education because education in the U.S. has included the idea of scientific literacy[1] as a central goal (cf. American Association for the Advancement of Science [AAAS], 1990; U.S. Congress, 1994). DeBoer (1997) traced the roots of scientific literacy back to the middle of the 19th century. In a review of the historical development of the idea of scientific literacy, DeBoer identified two different notions of scientific literacy: First, the general public should obtain an understanding of science "for its cultural value" (p. 76); and second, scientific literacy should also embrace a civic, democratic component. In his review, Roberts (2007) described these notions as two different components of the same idea, but each with a different focus: The first notion focuses on science subject matter (which Roberts termed 'Vision I'), whereas the second notion focuses on the interplay between scientific literacy and human affairs ('Vision II'). To take it to an extreme, Vision I would mean educating students to become highly specified scientists while Vision II would connote providing students with a sound scientific basis to become responsible citizens. Over the course of the past decades, the focus moved from what Roberts termed Vision I towards what he termed Vision II of scientific literacy. As early as 1969, Klopfer pointed out that more than "90 percent of all working people are engaged in occupations that are not directly related to science" (p. 201). Accordingly, Klopfer (1969) suggested introducing curricula for a "Scientific Literacy stream" (p. 204; corresponding to Vision II) in addition to those for a "Prospective Scientists stream" (p. 203; corresponding to Vision I). With the 'Science for All Americans' program (AAAS, 1990) and the Educate America Act (U.S. Congress, 1994) the idea of scientific literacy based on Vision II found its way into the National Science Education Standards of the United States (National Research Council [NRC], 1996).

Scientific literacy received international attention by being one particular assessment focus of the Third International Mathematics and Science Study (TIMSS) and of the Programme for International Student Assessment (PISA). By focusing on the question of what is "important for citizens to know, value, and be able to do in situations involving science and technology" (Organisation for Economic Cooperation and Development

---

[1] There is a particular discussion on the commonalities and differences between "scientific literacy" and "science literacy" in literature (cf. Roberts, 2007). Here, the expression "scientific literacy" is used exclusively to keep in line with Roberts's wording.

[OECD], 2006, p. 20), scientific literacy in PISA was distanced from the "mastery of all scientific knowledge" (p. 21). Accordingly, the society-oriented perspective of scientific literacy (corresponding to Vision II) was emphasized. Nevertheless, science subject matter should certainly be an indispensable part of scientific literacy, even if the contemporary, common emphasis is on the broad, society-oriented aspect of scientific literacy (Roberts, 2007).

Models of scientific literacy have been proposed to delineate this term. For example, a popular framework has been provided by Bybee (1997), who proposed a threshold model for scientific literacy consisting of four levels: (1) Nominal scientific literacy; (2) functional scientific literacy; (3) conceptual and procedural scientific literacy; and (4) multidimensional scientific literacy. Bybee's level of nominal scientific literacy refers to an understanding of scientific terms. However, it is characterized by the holding of misunderstandings concerning these terms. On the level of functional scientific literacy, scientific terms are used adequately through "Memorizing lists of scientific [...] vocabulary in textbooks" (Bybee, 1997, p. 57), but the use of terms is still limited to a small range of situations. The level of conceptual and procedural scientific literacy is related to an encompassing picture of scientific knowledge as a whole and an understanding of how single parts contribute to the entire discipline. This level also contains aspects concerning scientific procedures that are represented by "abilities of [... and] understandings about scientific inquiry" (Bybee, 1997, p. 58.), as described by the National Science Education Standards (NRC, 1996). The highest, so-called multidimensional level of scientific literacy is characterized by its broad scope. This level refers to "perspectives of science [...] that include history of scientific ideas, the nature of science [...], and the role of science and technology in personal life and society" (p. 61). People who do not reach any of these four levels of scientific literacy are classified as scientifically illiterate. Bybee assumed that this happens "because of their age, stage of development, or developmental disability" (p. 56).

Bybee's model of scientific literacy (Bybee, 1997) influenced the model of scientific literacy used in the PISA science framework (e.g., Prenzel, Schöps, Rönnebeck, Senkbeil, Walter, Carstensen, et al., 2007). However, in contrast to Bybee's model, the PISA science framework excluded scientific illiteracy; rather, "a continuum from less developed to more developed scientific literacy" (OECD, 2006, p. 25) was assumed. In doing so, scientific literacy referred to "an individual's:

- Scientific knowledge and use of that knowledge to identify questions, acquire new knowledge, explain scientific phenomena and draw evidence-based conclusions about science-related issues
- Understanding of the characteristic features of science as a form of human knowledge and enquiry

- Awareness of how science and technology shape our material, intellectual, and cultural environments
- Willingness to engage in science-related issues and with the ideas of science, as a reflective citizen" (OECD, 2006, p. 23).

In PISA, scientific literacy was operationalized through four elements (see Figure 1): scientific contexts, scientific competences, scientific knowledge, and attitudes towards science. Scientific competences were modeled to be influenced by knowledge and attitudes. Contexts created the relation to everyday life within our society.

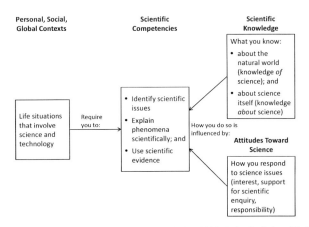

Figure 1. Framework for PISA 2006 science assessment (OECD, 2006, design by Bybee, McCrae, & Laurie, 2009).

Both of the described models – Bybee's threshold model (Bybee, 1997) and the PISA 2006 framework of scientific literacy (OECD, 2006) – contain aspects that refer to the role of science in our daily lives. Consequently, these models relate to Roberts's Vision II of scientific literacy (Roberts, 2007). However, Bybee elaborated a model of scientific literacy from only the theortical point of view, without suggesting means of measuring it. The PISA framework on the other hand was mainly intended to be a basis for developing test items. It thus represents an assessment-oriented approach.

Allgemeinbildung

As illustrated above, the idea of scientific literacy plays a central role in education in the United States. In Germany, education is strongly influenced by the idea of *Bildung*, in which science plays a secondary role only. However, following Germany's participation in the TIMS study, a growing debate on the performance of the educational system and its ideational foundation developed (cf. Klieme et al., 2004). Already in 1991 – and independent from these debates – Klafki suggested a revision of the classical notion of *Bildung* to meet contemporary educational challegenges (Klafki, 2007). With his work, Klafki aimed to contribute to the "development of a conception of *Bildung* that is oriented to the presence and future" (p. 15, translation and emphasis by the author). Klafki introduced a notion of *Allgemeinbildung* (general *Bildung*) that covered three aspects: 1) *Allgemeinildung* is an educational goal for everybody, 2) *Allgemeinbildung* is related to problems relevant for the general public and 3) *Allgemeinbildung* covers all "basic dimensions of human interest and abilities" (p. 54, translation by the author), such as cognitive, social, ethical, political or aesthetic aspects. The centerpiece of Klafki's notion of *Allgemeinbildung* is the idea of 'key problems typical of the era people live in' (*epochaltypische Schlüsselprobleme*). *Allgemeinbildung* means to be aware of such key problems and to perceive that everybody is jointly responsible for them and to be willing to solve them (Klafki, 2007). Klafki suggested the environmental problem as an example for such a key problem. With this example at hand, Klafki pointed out that through education, people need to be informed about scientific developments, and that they have to discuss and reach decisions about them.

In the scope of the debate about educational outcomes following the publication of the TIMSS and PISA results, Baumert (2002) proposed another notion of *Allgemeinbildung*. In this notion, *Allgemeinbildung* is related to the question of which objects of education are shared by societies across the world. Baumert suggested that those shared objects are represented by canonical knowledge and basic competences. According to Baumert, humans encounter the world and face problems from different perspectives. These perspectives are termed "modes of encountering the world" (p. 106; translation by the author according to Klieme et al., 2004). Four modes are differentiated: 'cognitive-instrumental', 'aesthetic-expressive', 'normative-evaluative' as well as 'problems of constitutive rationality'. These modes are thought to constitute the structure of canonical knowledge. Scientific knowledge is considered to be part of the cognitive-instrumental mode as such knowledge is required to encounter the world from a science perspective. According to Baumert, basic competences are required to enable individuals to make use of canonical knowledge. These competences are: 'mastery of the lingua franca', 'mathematisational competence', 'foreign-language competence', 'IT competence', and 'self-regulation of knowledge acquisition'. Baumert argued that his notion of *Bildung*, the modes of encountering the world and the basic competences, is relevant for all modern societies. Even if single aspects were subject to cultural differences, the framework as such can be found in standard and curricula documents of all modern societies.

Whereas the classical notion of *Bildung* was related to the formation of elite citizens (cf. Tenorth, 1994), the described theories of *Allgemeinbildung* are characterized by a focus on "the kind of non-vocational education that *every* person should engage in" (Biesta, 2002, p. 379, emphasis added). This means that these modern theories originate from the idea that *everybody* needs to responsibly participate in modern societies and cultures, and thus, *everybody* has to be prepared – that is, educated – for such participation.

Klafki's and Baumert's notions of *Allgemeinbildung* describe this idea based on a general level. How and to what extent the single disciplines contribute to *Allgemeinbildung* is not clearly defined. In contrast, the German education standards, which were introduced in 2004 and 2005, provide a more detailed framework that defines specific disciplines' contributions to *Allgemeinbildung*. The standards formulate competences that students are expected to have achieved through compulsory schooling. There are standards for German, English as a foreign language, mathematics (KMK 2004a, 2004b, 2004c), as well as biology, chemistry, and physics (KMK, 2005a, 2005b, 2005c). Within the science education standards, *Naturwissenschaftliche Bildung*[2] (scientific *Bildung*) is viewed as the overarching goal of science education (KMK, 2005a, 2005b, 2005c). Based on this educational goal, the science education standards define how science education contributes to *Allgemeinbildung*: "*Naturwissenschaftliche Bildung* enables individuals to actively participate in social communication and opinion formation on the development of technology and on scientific research; therefore, *Naturwissenschaftliche Bildung* is an essential part of *Allgemeinbildung*" (cf. KMK, 2005a, 2005b, 2005c, p. 6; translation and emphasis by the author).

In this context, *Naturwissenschaftliche Bildung* might be understood similarly to scientific literacy (Vision II). *Naturwissenschaftliche Grundbildung* carries central characteristics of scientific literacy (in the meaning of the U.S. research tradition): Students shall experience scientific phenomena; they shall learn about the language and history of science as well as about the methods and limits of scientific inquiry; and they shall be educated to communicate scientific information (KMK, 2005a, 2005b, 2005c). However, *Naturwissenschaftliche Bildung* differs from scientific literacy to some extent: For instance, *Naturwissenschaftliche Bildung* refers to *Allgemeinbildung*, while scientific literacy does not; furthermore, particular topics being included in scientific literacy (e.g., the nature of science) are not explicitly included *Naturwissenschaftliche Bildung* as proposed by the standards.

Given that *Allgemeinbildung* focuses on a preparation for participation in society, students need to be prepared for encountering the problems and issues of the current era (cf. Klafki's key problems). Because everyday life today is strongly influenced by science and technology, such problems will be related to science – at least to some extent. Referring to Klafki's notion of key problems, scientific literacy then is an essential prerequisite

---

[2] In the standards two terms are used: *Naturwissenschaftliche Bildung* (scientific *Bildung*) and *Naturwissenschaftliche Grundbildung* (scientific literacy). Based on the standards, it is unclear if these terms carry different meanings and how they are related.

for handling such key problems: Citizens of this current era will have to make decisions on problems such as the environmental issue – they will have to decide to what extent this issue is relevant to today's society and how it should be tackled. Likewise, Thomas and Durant (1987) pointed out that a public understanding of science (i.e., scientific literacy) yields "[b]enefits to [d]emocratic [g]overnement" (p. 5). In democratic societies citizens are involved in political decision-making. A particular amount of these decisions to be made are science-related because much of scientific research is publicly funded, and scientific research results influence our private and societal lives (Thomas & Durant, 1987). Scientific literacy is therefore needed to "promote more democratic decision-making [... and] more effective decision-making" (Thomas & Durant, 1987, p. 5).

In summary, the classical notion of Bildung has been replaced by a modernized notion of *Allgemeinbildung* that focuses on the advancement of the society. The educational goal set by the German science education standards, *Naturwissenschaftliche Bildung*, shall contribute to this advancement. Likewise, scientific literacy was shown as having an important role in science-related, social issues, and thus in terms of *Allgemeinbildung*. Interestingly, this relevance does not become evident from the theories of *Allgemeinbildung*: Klafki (2007) does not explicitly emphasize the necessity of a sound science education, and thus scientific literacy, to handle key problems of our era. Likewise, scientific competence is not considered a basic cultural tool according to Baumert (2002). At the very least, however, the introduction of *science* standards as well as the focus on *Naturwissenschaftliche Bildung* signified a stronger focus on science education and scientific literacy in the German educational system[3].

Allgemeinbildung and the nature of science

One element of scientific literacy is an adequate understanding about the nature of science (cf. Bybee, 1997). The term 'nature of science' typically refers to the characteristics of science; that is, of scientific processes and scientific knowledge. Understanding these characteristics is consistently viewed as an important goal of science education (e.g., Driver, et al., 1996; McComas, Clough, & Almazroa, 1998; Lederman, 2007). Such understanding is of particular importance with respect to socioscientific discussions as Driver et al. (1996) illustrated. According to these authors, socioscientific issues sometimes involve controversial scientific facts. Such disagreements might be induced by issues as a "dispute about how the laboratory findings relate to the complex and messy real-world situation" (Driver et al., 1996 p. 18), or by uncertainties as to what extent knowledge can be reliably applied in a new context. Driver et al. pointed out that understanding the nature of science helps to understand and evaluate such controversial scientific facts.

---

[3] However, the German education standards were first introduced for German, English and mathematics, and for the science disciplines only some time later.

Likewise, a case study by Millar and Wynne (1988) illustrates why an understanding of the nature of science matters in socioscientific contexts. These authors vividly illustrated the need for a sufficient view of the nature of science in order to adequately understand science-related media reports. In particular, media reports on the Chernobyl disaster and its implications for the British population were evaluated. The media reports were shown to reflect the uncertain nature of scientific knowledge. According to Millar and Wynne, science experts provided statements about the intensity and development of radiation for particular British areas and about possible aftermath (e.g., death due to cancer). Like all scientific knowledge, these statements were not absolute, but tentative. Yet, there were people who expected unequivocal statements by the science experts, since they were under the impression that science produces absolute and true facts. Thus, in the case of the Chernobyl disaster and aftermath, citizens would have needed to understand the characteristics of scientific knowledge and the processes of how the knowledge was produced to have been able to responsibly evaluate information provided by science experts.

In the same way, understanding about the nature of science enables citizens to become involved with key problems of the current era; the Chernobyl disaster and its implications might even be understood as such a key problem. Consequently, an adequate understanding about the nature of science contributes to *Allgemeinbildung* sensu Klafki (2007). Citizens need to be able to use scientific information to form an opinion about such issues. To be able to do so, the average citizen needs to understand, for example, how scientific information is created or what characteristics scientific information carries. Likewise, an understanding about the nature of science may be considered an important aspect of Baumert's (2002) notion of *Allgemeinbildung*. In this notion, the cognitive-instrumental mode may be understood as understanding the nature of science because such understanding enables citizens to understand both the power and limitations of scientific approaches. This enables people to choose the appropriate mode(s) to encounter the world for a particular problem: Sometimes, the aesthetic-expressive or normative-evaluative modes need to be employed instead of blindly believing in the significance of science; sometimes, a scientific approach might help to solve a problem. In summary, an adequate understanding about the nature of science may be considered an important aspect of *Allgemeinbildung*, in the senses of both Klafki and Baumert.

Due to the importance of an understanding about the nature of science with respect to *Allgemeinbildung*, it does not come as a surprise that German science education researchers agree on the central role of an understanding about the nature of science for education (cf. Höttecke, 2001a; Kircher, 2007, Urhahne, Kremer, & Mayer, 2008). However, understanding the nature of science is not considered within the concept of *Naturwissenschaftliche Bildung*. Therefore, it is not included in the science education standards and thus, it is not obligatory for all German students in secondary schooling up to grade 10. This imbalance is unreasonable. According to Klieme et al. (2004), the advancement of future citizens as one central educational goal should underlie the German education standards. An adequate understanding of the nature of science was shown above as important to include in such education. Consequently, the German science education standards (KMK, 2005a, 2005b, 2005c) currently leave a blank with respect to the educational goal demanded by Klieme et al. The present work, therefore, aims to draw an approach for how the standards – for non-vocational education – can be expanded to include the concept of nature of science.

2   Competence

Prior to the introduction of national education standards, the German education system was input-based, meaning it was controlled by curricula or requirements for teacher training, for example; the standards, however, define students' achievement (cf. Helmke, 2000; Klieme et al., 2004). Instead of providing "lists of content and material to make educational goals concrete" (Klieme et al., 2004, p. 17), as was the case in the former input-based system, the recently introduced standards define learning outcomes by sets of competences students should achieve by the end of compulsory education. A detailed definition of the concept of competence is given in the first section of this chapter. In their expertise on educational standards, Klieme et al. (2004) also pointed out the need to have competence models in order to specify educational goals set by the standards. Accordingly, a broad research program on developing such models of competence came up in educational research, and in science education research, in particular. The second section gives an overview of this research.

2.1 Concepts of Competence

Klieme and Hartig (2007, see also Klieme, Hartig, & Rauch, 2008) have traced the term of competence back to the 1950s. Three different approaches to outlining the idea of competences were extracted from a literature review: a generic, a normative, and a functional-pragmatic approach (Klieme & Hartig, 2007). These three approaches each represent a particular conception of the same term of competences.

The 'generic approach' is based on Noam Chomsky's work concerning the collective phenomenon of language acquirement (Chomsky, 1986). "Chomsky understood linguistic competence as a universal, inherited, modularized ability to acquire the mother tongue" (Weinert, 2001a, p. 47). Characteristic to Chomsky's theory is the differentiation between competence and performance (e.g., Klieme & Hartig, 2007, Klieme et al., 2008; Weinert, 2001a). This linguistic competence is understood as a system generating performance which differs between individuals and between situations. Competence, as understood by Chomsky, is seen to be a trait "common to all human beings [… and] interindividual differences only relate to *performance* as the actual realization of a competence" (Klieme et al., 2008, p. 5, italics original).

The 'normative approach' is mainly connected with the work of the educational philosopher Heinrich Roth (e.g. Roth, 1971). Roth combined the idea of competence with the idea of *Bildung*. As discussed in Chapter 1, the classical notion of *Bildung* is connected with the self-formation of humans. One indispensable part of this self-formation is maturity (*Mündigkeit*), which became relevant especially in the era of enlightenment (cf. Kant's definition of enlightenment being "man's release from his self-incurred tutelage", 1992, p. 90). Roth (1971) seizes this idea of maturity and connects it to the concept of competence. In his view, "maturity [*Mündigkeit*] should be interpreted as competence in a threefold sense: a) as self-competence – the ability to be responsible for your own action, b) professional competence – the ability to act and judge in a particular profession, and hold responsible, c) social competence – the ability to act and judge, and hold responsible, in professional or social areas that are relevant in social, societal or political terms" (Roth, 1971, p. 180, as cited by Klieme et al., 2008, p. 6, italics and brackets original). Roth's idea of competence still maintains the broadness typical of the notion of *Bildung*. In an attempt to integrate *Bildung* with the idea of outcome-based instruction, Baumert (2002) used the concept of competence to define basic cultural tools as a component of *Allgemeinbildung*. Similarly, the ability to successfully and responsibly tackle key problems within Klafki's notion of *Bildung* (Klafki, 2007) might also be interpreted as competence according to Roth's normative approach. Altogether, Roth's and Baumert's notions of competence as well as the abilities defined by Klafki refer to ideals and represent an all-embracing encountering of one's life and the world. At first glance, competences focusing on applying knowledge in particular situations seem to be incompatible with *Bildung*, which was traditionally connoted with the learning of a large amount of knowledge and the noble aim of self-formation. Roth's and Baumert's works demonstrate that competence and *Bildung* are not as incompatible as they seem to be and, in fact, can be combined (see also Tenorth, 2003).

The "functional-pragmatic" approach, to which most of current science education research projects refer, is characterized by a much narrower, situation-related focus – in contrast to the generic and the normative approaches (Klieme & Hartig, 2007). This pragmatic approach is traced back to David McClelland, who criticized the trend of broadly measuring students' general cognitive abilities by showing that these abilities were not powerful predictors for students' future professional success (McClelland, 1973). Instead, "testing for competence rather than 'intelligence'" was suggested (McClelland, 1973, p. 1). A remarkable feature of competences according to McClelland is their close relation to particular work and activities. Thus, "any kind of individual attribute [e.g., cognitive abilities, personal factors, etc.] may be perceived as 'competence' as far as it serves to predict success in concrete achievement" (Klieme et al., 2008, p. 7).

On behalf of the OECD, Weinert reviewed the literature related to the construct of competence, identified nine notions of competence and discussed their theoretical and practical appropriateness for educational contexts (Weinert, 2001a; Klieme & Hartig, 2007). For Weinert, "it seems theoretically and practically expedient to restrict the concept of competence to domain-specific learning and domain-specific skills, knowledge, and strategies" (Weinert, 2001a, p. 57). This illustrates the emphasis Weinert puts on the domain-specificity of competences; another characteristic feature of competences is that competences can be acquired in contrast to inborn cognitive abilities such as intelligence (Klieme & Leutner, 2006). Moreover, Weinert suggests that motivational aspects be included in a definition of competence. Weinert (2001b) later defines competences as "abilities and skills, which individuals have available or can learn, in addition to motivational, volitional and social disposition and abilities to successfully and responsibly use the solutions of problems in variable situations" (p. 27f, translation by the author). This shows that according to Weinert (2001a, 2001b; cf. Klieme, & Hartig, 2007) domain specificity is a central characteristic of competence, which includes motivation and interest for a specific problem and/or a certain domain. Consequently, similar to McClelland's notion of competence, Weinert's notion corresponds to the pragmatic approach. Although Weinert's definition of competences and the notion of competence utilized in PISA (e.g. OECD, 1999, 2006) are not exactly the same, there are parallels between the two: "At the heart of the PISA 2006 definition of scientific literacy [lie competences which] require students to demonstrate, on the one hand, knowledge, cognitive abilities, and on the other, attitudes, values and motivations as they meet and respond to science related issues" (OECD, 2006, p. 20). This demonstrates the involvement of both cognitive and motivational aspects similar to Weinert's definition.

Weinert's (2001b) definition became the basis for current science education standards (KMK 2005a, 2005b, 2005c). Consequently, the project to benchmark science education standards in Germany, the *Evaluation der Standards in den Naturwissenschaften für die Sekundarstufe I*[4] project (ESNaS; cf. Kauertz, Fischer, Mayer, Sumfleth, & Walpuski, 2010), was based on Weinert's definition; and so are other projects that are concerned with the assessment of students' competences and that refer to the science education standards (e.g., Bernholt, 2010; Einhaus, 2007; Hammann, 2004; Kulgemeyer & Schecker, 2009; Schmidt, 2008). Even if Weinert's definition of competence includes both cognitive and motivational aspects, Weinert (2001a) suggested modeling and assess-

---

[4] Evaluation of the National Educational Standards for Natural Sciences at the Lower Secondary Level

ing these aspects separately from each other. Accordingly, Hartig and Klieme (2006) along with Klieme and Leutner (2006) proposed focusing on cognitive aspects when dealing with competence assessment; therefore, the notion of competence was condensed into cognitive, learnable, and domain-specific features. Additionally, Hartig and Klieme pointed out that the internal structure of competences results from particular situations and problems. This narrower concept of cognitive competence represents the core of a recently established priority program of the German research foundation (cf. Klieme & Leutner, 2006).

In summary, among the various approaches to define and describe the idea of competences, a quite narrow definition became widely accepted. This definition is particularly appropriate for the assessment of competences. Core features of this definition of competences are their (a) cognitive, (b) learnable, and (c) domain-specific aspects.

2.2 Competence Models

With an increasing focus on competences, modeling and assessing competence(s) has gained central importance in German educational research in various domains (Koeppen, Hartig, Klieme, & Leutner, 2008; Prenzel, Gogolin, & Krüger, 2007). For instance, there is research on competence models with respect to deductive reasoning (Spiel & Glück, 2008); self-regulated learning (Wirth & Leutner, 2008); languages like German and English (Jude, Klieme, Eichler, Lehmann, Nold, Schröder, et al., 2008); mathematics (Reiss, Heinze, & Pekrun, 2007); and science (e.g., Bernholt, 2010; Hammann, 2004; Kauertz, 2008). Modeling and assessing of competences are closely related to each other. Koeppen et al. (2008) emphasized that "Valid measures of competence need to be based on theoretically sound and empirically tested competence models." (p. 63). Similarly, Schecker and Parchmann (2006) pointed out that the description of competences requires competence models, which should be additionally appropriate for the assessment (and learning) of competences. In summary, competence models represent the theoretical basis for theory-based research in the field of competence assessment. Only on the basis of a theoretical competence model can assessment outcomes be meaningfully interpreted.

Schecker and Parchmann (2006) proposed the following categorization of competence models. In terms of the purpose of assessments, normative and descriptive models were distinguished. In the authors' view, normative models define required competences learners are expected to meet (a priori). In contrast, descriptive models are used to portray the observed competence of learners (post hoc). In terms of the period of evaluation, Schecker and Parchmann differentiated between structural and developmental models. According to these authors, structural models refer to a particular point in time during schooling (e.g., the end of grade 10), while developmental models aim at describing the change in students' competence over a certain period of time (e.g., from kindergarten through compulsory school). According to Schecker and Parchmann, German science education standards (KMK, 2005a, 2005b, 2005c) and Bybee's (1997) model of scientific literacy are examples for normative, structural models of competence.

In fact, scientific literacy and scientific competence are closely related to each other. Basically, being scientifically literate is equivalent to being scientifically competent. In other words, being scientifically literate means having scientific competence at one's disposal. Bybee's (1997) model of scientific literacy provided a basis upon which later competence models built. For example, based on the results of the TIMS and PISA studies, items were categorized into groups related to Bybee's levels of scientific literacy (so-called 'post hoc models'; cf. Neumann, Kauertz, Lau, Notarp, & Fischer, 2007). These post-hoc models, however, have not been immune to criticism. One point of criticism was that only those competences represented by the test items were included in the models (cf. Schecker & Parchmann, 2006). This does not necessarily mean that all levels of Bybee's model have been adequately included. Another point of criticism refered to items not being sufficiently classified with respect to Bybee's levels, and thus brought into question the models' empirical validity (cf. Neumann et al., 2007). In contrast, so-called 'a priori models' are developed based on a theoretical background and are operationalized into test items *before* gathering empirical data. A priori models are seen to more effectively address such criticism. Accordingly, several a priori models have been proposed over the last decade.

German science education standards represent such an a priori model of competence: On one dimension, the standards differentiate areas of competence (use of content knowledge, acquirement of knowledge, evaluation, and communication) that categorize subsets of competences; on the other dimension, they define three mastery levels (reproducing, applying, and transferring). However, nobody in science education research seems to having attempted to validate this two-dimensional model ever.

One of the first models operationalizing the standards was proposed by Schecker and Parchmann (2006). These authors developed a five-dimensional model of competence (*Bremen-Oldenburger Kompetenzmodell*, BOlKo) which was suggested to serve as a starting point for evaluating the standards concerning all three science disciplines. Because it consisted of five dimensions, this model was very complex. Researchers have attempted to validate only parts of the BOlKo, therefore (e.g., Einhaus, 2007; Schmidt, 2008). Furthermore, in these studies the assumed structure of the investigated dimensions has not been able to be corroborated (cf. Einhaus, 2007; Schmidt, 2008).

Parallel to Schecker and Parchmann's work, Neumann, et al. (2007; see also Kauertz, 2008) suggested a model to describe the structure and the development of physics competence based on the idea of vertical linkage in the context of learning (Fischer,

Glemnitz, Kauertz, & Sumfleth, 2007). At the core of this model was the idea of complexity of information: Kauertz (2008) demonstrated that test items become more difficult the more complex information there is to be processed (see also Kauertz & Fischer, 2006). Moreover, 29% of variance in item difficulty was due to the different complexity levels (Kauertz, 2008). As a consequence, complexity provided a criterion to define levels of mastery in a competence model. Similarly, Bernholt, Parchmann and Commons (2009) adapted the so-called 'Model of Hierarchical Complexity' (Commons, Trudeau, Stein, Richards, & Krause, 1998) to describe competence regarding the use of chemistry content knowledge. Bernholt et al. (2009) were able to identify a considerable influence of the complexity level on item difficulty. Bernholt (2010) reported that 56% of variance in item difficulty was explained by the proposed complexity dimension. However, Bernholt's (2010) investigations refer to only a small amount of basic science concepts; on the other hand, Kauertz's (2008) findings correspond to a wide range of basic concepts of scientific knowledge.

Since Kauertz's findings (2008) were seen as likely to be valid for other areas of content, too, the nationwide project ESNaS (*Evaluation der Standards in den Naturwissenschaften für die Sekundarstufe I*), which evaluates German science educational standards (KMK, 2005a, 2005b, 2005c), used this work as well as that by Neumann et al. (2007) as a starting point. Based on Kauertz's findings, a three-dimensional model of competence was developed (Kauertz et al., 2010). This model was used to describe students' competence concerning all three science subjects taught in German schools (biology, chemistry, and physics). One dimension of the model contains four areas of competence. The 'use of content knowledge' is related to an understanding of the central concepts of the sciences – for instance, the concept of energy, the concept of chemical reactions, and the concept of individual and evolutionary development. The 'acquirement of knowledge' comprises inquiry-related competences that are mainly related to processes, like being able to formulate appropriate hypotheses or to plan and conduct investigations. The area of 'evaluation' includes the ability to assess socioscientific issues (e.g., from ethical or economic perspectives). Finally, 'communication' corresponds to the ability to communicate scientific information in a way that is appropriate for the audience and the content (Kauertz et al., 2010). A second model dimension describes the type of information that has to be processed by a student when solving a problem. This dimension was also based on the idea of complexity as described by Kauertz (2008). Altogether, five levels of complexity for content related competences were differentiated: one fact (Level I), two facts (Level II), one relation (Level III), two relations (Level IV), and overarching concept (Level V). Facts or level I features were defined as smallest unit of the respective area of competence. In accordance with Kauertz's findings (2008), the five levels of complexity were assumed to be ordered hierarchically (increasing from I to V) and to generate test item difficulty. The third dimension is composed of four cognitive processes. This dimension describes how information has to be cognitively processed when solving a given problem. A distinction is drawn between reproducing, selecting, organizing, and integrating information. As in the case of complexity, cognitive processes were assumed to be ordered hierarchically (increasing from reproducing to integrating) and to have an impact on the difficulty of test items. The model is visualized in Figure 2. Model of competence concerning science as it is used in the project ESNaS, being referred to as 'ESNaS-model' (Kauertz et al., 2010, translation by the author). and represents the basis for the development of items, which will be employed to determine German students' competence and to subsequently benchmark science education standards (Walpuski, Kampa, Kauertz, & Wellnitz, 2008). The validation study of this three-dimensional ESNaS-model is currently in progress. However, first results indicate that the proposed model characteristics are corroborated by empirical data. In particular, the dimensions of *cognitive processes* and *complexity* meet the expected influence on difficulty of test items on the use of content knowledge for biology, chemistry and physics (cf. Kauertz et al., 2010).

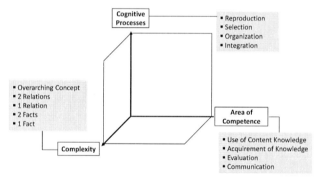

Figure 2. Model of competence concerning science as it is used in the project ESNaS, being referred to as 'ESNaS-model' (Kauertz et al., 2010, translation by the author).

3 Nature of Scientific Inquiry (NOSI) and Nature of Scientific Knowledge (NOS)

The four areas of competence considered in the German science education standards do not include aspects of nature of science, whereas standards and educational documents of several other countries do (cf. McComas & Olson, 1998). The topic 'nature of science' involves aspects of history, epistemology, and sociology of science (cf. Kircher, 2007; Lederman, Abd-El-Khalick, Bell, & Schwartz, 2002). As such, nature of science is in principle related to the discipline of philosophy. In the first section, a literature review provides an overview of why and which aspects of this philosophy-related topic should reasonably be included in science education. In this section, moreover, two central components of nature of science are identified: Nature of Scientific Inquiry and Nature of Scientific Knowledge. An overview of current research on conceptions concerning nature of science will then be provided in the second section.

3.1 Rationale and Delineation

Science education researchers widely agree that there is no consensus regarding a definition of nature of science in philosophical literature (e.g., Alters, 1997; Driver et al., 1996; Lederman, 2006; McComas et al., 1998; Osborne, Collins, Ratcliffe, Millar, & Duschl, 2003). This does not come as a surprise, given the different approaches that can be found among the most important philosophers of science like Karl R. Popper, Thomas S. Kuhn, Imre Lakatos and Paul Feyerabend (a detailed description of the diversity of philosophy of science can be found with Chalmers, 2007). Popper's idea of falsification-ism was based on the assumption that no scientific theory can be experimentally proven to be correct, but that a theory must still be refutable (Popper, 1976). Kuhn (1976) argued that such falsificationism cannot be found in science history; but that instead, science can be described by paradigms developing in cycles of normal science, crisis induced by puzzling and intractable anomalies, and following scientific revolutions. A similar description was proposed by Lakatos (1978). He introduced research programs made up by a hard core of theories and a number of auxiliary hypotheses protecting the hard core against falsification (Lakatos, 1978). Popper, Kuhn and Lakatos aimed at describing what makes science unique. In contrast, Feyerabend (1993) provocatively stated that there is no scientific method by which scientific progress can be described, and, thus, pleaded "against method" (p. 14) and for a principle of "anything goes" (p. 14) instead. The above examples show that there is no consensus on the characteristics of the nature of science among philosophers. Despite this diversity of philosophical definitions concerning the nature of science, science education researchers do agree on two points: First, nature of science is an indispensable part of science education in school; and second, there is only little disagreement about which aspects of nature of science are relevant for a school science curriculum (Lederman, 2007; McComas et al., 1998; Osborne et al., 2003).

Understanding of the nature of science as a goal of science education

Driver et al. (1996) provided a thorough rationale and five arguments for why nature of science should be included in science education to promote scientific literacy:

- "an understanding of the nature of science is necessary if people are to make sense of the science and manage the technological objects and processes they encounter in everyday life" *utilitarian argument* (p. 16)

- "an understanding of the nature of science is necessary if people are to make sense of socioscientific issues and participate in the decision-making process" *democratic argument* (p. 18)
- "an understanding of the nature of science is necessary in order to appreciate science as a major element of contemporary culture" *cultural argument* (p. 19)
- "learning about the nature of science can help develop awareness of the nature of science, and in particular the norms of the scientific community, embodying moral commitments which are of general value" *moral argument* (p. 19)
- "an understanding of the nature of science supports successful learning of science content" *science learning argument* (p. 20)

Concerning scientific literacy, Bybee (1997) also included nature of science in his model. Nature of science aspects were located on the multidimensional level (see Chapter 1) which "is [the] one that many citizens experience" (p. 61). Likewise, aspects of nature of science can be found in science education documents across the world (see McComas and Olson's review of education documents from five countries in 1998). The Benchmarks for Science Literacy (AAAS, 1993) divide nature of science into three topics. First, students should learn about the scientific worldview, which is characterized by the endeavor to "figure out how the world works" (AAAS, 1993, p. 5). This aspect also contains the insight that rules that have been explored concerning one part of the universe are often appropriate for another part and that the knowledge gained "is both stable and subject to change" (p. 5). The second aspect refers to scientific inquiry as a complex process; this process includes several different methods and approaches: the influence of creativity and imagination, the necessary condition of logical consistency and empirical evidence, and the process of justification in the scientific community. Third, students should be aware that science is an enterprise that humankind has been concerned with for a long time. Because science is conducted by humans, it carries social aspects from two different perspectives. On the one hand, science has a social component itself; on the other hand, it is related to issues that are relevant for the non-scientific society.

In accordance with the Benchmarks for Science Literacy (AAAS, 1993), the National Science Education Standards (NRC, 1996) include categories of content standards pertaining to both the 'history and nature of science' and 'scientific inquiry'. The history and nature of science set covers the aspect "science as a human endeavor" (p. 200) corresponding to social structures and ethics of the scientific community. Additionally, the aspect "nature of scientific knowledge" (p. 201) addresses characteristics such as the fact that scientific knowledge is empirical. Finally, the history and nature of science set includes "historical perspectives" (p. 201) on science, which include illustrating the change to which scientific knowledge is subject and exemplifying that different cultures have always contributed to scientific progress (and still do). The set of content standards on 'scientific inquiry' makes a clear distinction between the *ability to conduct* scientific inquiry and an *understanding about* scientific inquiry. Abilities to conduct scientific inquiry embrace skills related to scientific investigations, like hypothesizing, conducting investigations or analyzing data. In contrast, an understanding about scientific inquiry corresponds to a sort of meta-knowledge on the process of inquiry, including its conditions or its interrelation to technology and mathematics. Thus, this understanding partly overlaps with the aspect of "science as a human endeavor". Accordingly, one could also argue in favor of including the understanding about scientific inquiry under the umbrella of nature of science.

Kircher (2007), a German science education researcher, pointed out that learning *about* science includes aspects of nature of science (aspects from epistemology and philosophy of science), but also ethical and social issues (see also Kircher & Dittmer, 2004). In particular, the interplay of science and society and the ethical dimension of science are embedded in German science education standards for secondary education up to grade 10. In contrast, the concept of nature of science as such is omitted. Nonetheless, physics and chemistry standards consider the promotion of a typical scientific worldview among students (KMK, 2005b, 2005c). Moreover, the standards for all three science disciplines emphasize the necessity of *Naturwissenschaftliche Bildung* as a basis for actively participating in discussions of the society and forming an opinion concerning scientific issues (KMK, 2005a, 2005b, 2005c). In arguing that an adequate view of the nature of science is a prerequisite for such participation (cf. Chapter 1), one might say that nature of science is already a part of the standards, yet only implicitly. However, German science education researchers agree that the nature of science plays a central role in school science education (Höttecke, 2001a; Kircher, 2007; Kircher & Dittmer, 2004; Schecker, Fischer, & Wiesner, 2004). With respect to science education in general, Kircher and Dittmer (2004), for instance, point out that – besides science content – aspects of epistemology and philosophy of science as well as ethical aspects of science need to be included in science lessons. With respect to (non-vocational) upper secondary school, the inclusion of nature of science aspects is even more strongly emphasized: For example, Schecker et al. (2004) argue that science propaedeutics – that is, the introduction into the discipline of science – is a major role of physics education in upper secondary school (grades 11 to 12/13). One indispensable part of propaedeutics is promoting an adequate view of the scientific discipline of physics by addressing nature of science aspects in physics instruction.

Approaches to outline a curriculum on the nature of science

As pointed out in the example of Popper, Kuhn, Lakatos, and Feyerabend, from the perspective of philosophy, there do exist different characterizations of what science is or how science works. From the perspective of science education however, three recent works have demonstrated that there is a particular consensus on which aspects of nature of science should be included in a curriculum on nature of science.

McComas and Olson (1998) analyzed eight science education documents from five countries – the United States, Australia, England/Wales, New Zealand and Canada – focusing on the presence of elements of nature of science. More than 25 statements related to philosophical, sociological, psychological and historical aspects of science were identified in these documents. Yet, not all statements occurred in all eight documents. Indeed, McComas and Olson pointed out that all documents have not been written with the aim to provide an all-encompassing picture of science content. Thus, important aspects may not appear within the majority of the documents even if they were indeed important. McComas and Olson therefore refrained to compile a list of aspects of nature of science in order of the degree of appearance in the analyzed documents even if it would have been possible based on their data.

Osborne et al. (2003) conducted a Delphi study involving science educators, leading scientists, historians, philosophers, sociologists of science and science communicators. These experts were asked about their opinion on what should be taught "(a) [...] about the methods of science [...,] (b) [...] about the nature of scientific knowledge [..., and] (c) [...] about the institutions and social practices of science" (p. 699). Osborne and colleagues were able to identify a consensus view concerning 10 themes. These findings were then aligned with "the most prevalent ideas about science" found by McComas and Olson (1998); the aligned findings are shown in Table 1. A comparison shows a considerable degree of agreement.

Lederman (2006; also Lederman et al., 2002; Lederman, 2007) focuses on the question of which aspects of nature of science should be emphasized in science research and curriculum development. To elaborate such aspects, Lederman used three criteria: (1) if an aspect is accessible to students, (2) if there is consensus about the aspect, and (3) if the aspect is useful for all citizens. Employing these criteria, seven relevant aspects of nature of science were identified which were in line with the suggestions of the Benchmarks for Science Literacy (AAAS, 1993) and the National Science Education Standards (NRC, 1996). Lederman moreover suggested to clearly distinguishing between nature of scientific knowledge and scientific inquiry (Lederman, 2006; Schwartz, Lederman, & Lederman, 2008). This distinction corresponds with the terminology used in PISA (OECD, 2006, p. 22, italics original): "*Knowledge about science* [which is to be distinguished from *knowledge of science*] refers to knowledge of the means (scientific enquiry) and goals (scientific explanations) of science." This quote illustrates the reasonable differentiation between knowledge about the process (scientific inquiry) and the product (scientific knowledge and explanations). The seven aspects of nature of science proposed by Lederman (2006, see Table 1) refer to the epistemology of science, in particular to the nature of scientific knowledge. For research tradition reasons, however, these aspects are subsumed under the term Nature of Science (NOS). Likewise, under the leadership of Norman G. Lederman eight statements concerning the understanding about scientific inquiry were extracted from the U.S. National Science Education Standards (personal communication, 15.10.2007; see Table 1). Schwartz et al. (2008), who seem to be the first to introduce the term 'Nature of Scientific Inquiry (NOSI)', outlined eight aspects based on science standard documents and science education research reports as well: "a) Questions guide investigations, b) multiple methods of scientific investigations, c) multiple purposes of scientific investigations, d) justification of scientific knowledge, e) recognition and handling of anomalous data, f) sources, roles of, and distinctions between data and evidence, and g) community of practice" (p. 4). Comparing this list with Lederman's eight statements (Table 1) reveals a high alignment and overlap. Therefore Lederman's eight statements will be referred to as NOSI in this thesis, as well.

Table 1 provides a juxtaposition of (1) the statements extracted from educational documents by McComas and Olson (1998) and selected by Osborne et al. (2003), (2) the consensus themes Osborne et al. (2003) found from their Delphi study, (3) aspects of NOS proposed by Lederman (2006), and (4) aspects of NOSI proposed by Lederman and his research group (personal communication, 15.10.2007). On a general level, Table 1 shows that aspects of NOS can be better aligned with the work by McComas and Olson (1998) and by Osborne et al. (2003) than the aspects of NOSI can be. For example, *Inquiry procedures are guided by the question asked* is related to *Creativity/Science and Questioning*, however they do not exactly match each other. This observation is most likely due to the missing differentiation between NOS and NOSI in both works by McComas and Olson (1998) as well as by Osborne et al. (2003). Some of the aspects (e.g., *Tentativeness* or *Creativity*) can be easily located in this comparison, since they already match in their wording. Other aspects only partially refer to each other. For example, *Scientific data are not the same as scientific evidence* relates quite to *Analysis and Interpretation of Data*, which also includes that "data do not 'speak by themselves'" (Osborne et al., 2003, p. 708). Likewise, the statement that *All scientists performing the same procedures may not get the same results* applies to the statement that data "'can be variously interpreted'" (ibid.). To a certain extent, *Research conclusions must be consistent with the data collected* refers to the reliance on empirical evidence as well. It is also an important criterion for *Critical Testing*. It is confusing that Osborne et al. include *Scientific Method* in their list, which would lead one to assume that it contradicts the NOSI statement that *There is no single scientific method*. When checking with Osborne et al. (2003) on the meaning of this theme, it becomes clear that critical testing is agreed to be the most central method in science, but not the only one – which also is mirrored by the theme *Diversity of Scientific Thinking*. In summary, the comparison illustrates that, on a

Table 1. Comparison of aspects concerning Nature of Science / Nature of Scientific Inquiry from different works of science education research. Left two columns are quoted from Osborne et al. (2003, p. 713).

McComas & Olson*	Osborne et al.**	Nature of Science†	Nature of Scientific Inquiry††
Scientific knowledge is tentative	Science and Certainty	Tentativeness	All scientists performing the same procedures may not get the same results./ Scientific data are not the same as scientific evidence.
Science relies on empirical evidence	Analysis and Interpretation of Data	Distinction between observation and inference/ Empirically based	Research conclusions must be consistent with the data collected
Scientists require replicability and truthful reporting	Scientific Method and Critical Testing		
Science is an attempt to explain phenomena	Hypothesis and Prediction		Scientific investigations all begin with a question, but do not necessarily test a hypothesis.
Scientists are creative	Creativity/Science and Questioning	Creativity	
Science is part of social tradition	Cooperation and collaboration in the development of scientific knowledge	Culturally embeddedness	
Science has played an important role in technology	Science and Technology		There is no single set and sequence of steps followed in all scientific investigations (i.e., there is no single scientific method).
Scientific ideas have been affected by their social and historical milieu	Historical development of scientific knowledge	Subjectivity (Theory-ladenness)	
	Diversity of Scientific Thinking		
Changes in science occur gradually			Inquiry procedures are guided by the question asked.
Science has global implications			Inquiry procedures can influence the results.
		Distinction between theory and law	
New knowledge must be reported clearly and openly			Explanations are developed from a combination of collected data and what is already known.

*McComas & Olson (2002); ** Osborne et al. (2003); † according to Lederman (2006, 2007); †† according to N. G. Lederman and his research group (personal communication, 15.10.2007) general level, there is agreement on which aspects of nature of science are relevant for science education in school.

3.2 Research on Students' Conceptions of the Nature of Science

The previous section explained that there is a consensus among science education researchers on nature of science as an important goal of science education. This section focuses on to which extent this goal has been reached, and, thus, on students' conceptions of nature of science. Lederman (2007) provided one of the most extensive and thorough reviews of research on nature of science over the past decades, in which three relevant issues related to students' conceptions could be identified: instruments to assess conceptions of nature of science, typical inadequate conceptions that have been identified and approaches to change students' conceptions regarding nature of science.

Instruments to assess conceptions of the nature of science

The earliest instrument, identified in Lederman's (2007) review, is the Science Attitude Questionnaire by Wilson (1954). This indicates the long tradition of science education research that is concerned with the assessment of nature of science conceptions. Most of the 28 instruments reviewed by Lederman are Likert-type questionnaires, which Lederman viewed as a consequence of the trend towards formal, easy-to-administer assessment methods that emerged in the 1960s (cf. Lederman, Wade, & Bell, 1998; Lederman, 2007). Although there is a vast amount of instruments on nature of science, Lederman's review shows that most of these instruments suffer from validity issues. Concerning validity, two main aspects are criticized. The first validity problem refers to the instruments' scope. Even if they were meant to assess conceptions of nature of science, many of the reviewed tests consist of items that go beyond the scope of nature of science rather than clearly focusing on it; such instruments are for example a) focusing on students' skills concerning scientific inquiry, b) not assessing *knowledge* concerning nature of science but rather *affective attitudes* towards nature of science, or c) missing emphasis on the epistemology of science for the benefit of including science as an institution (Lederman, 2007). The second validity problem is due to narrowing in on one specific philosophical view when scoring the test items. As discussed in Section 3.1, there are several accepted works by philosophy of science scholars, each with their own descriptions of science. If a test item includes debatable aspects with respect to the differences in description, the scoring of this item, and thus the interpretation of the test score is questionable and probably not valid (Lederman et al., 1998).

An attempt to face such validity issues is represented by the Views on Nature of Science (VNOS) questionnaire series by Lederman et al. (2002). These questionnaires show two main advantages. First, they were developed based on a thoroughly delineated construct – namely, Lederman's aspects of Nature of Science (Lederman et al., 2002).

Second, the questionnaires' items are open-ended. Lederman et al. argued that this answering format allows for an uninfluenced answer and supports a deep and detailed assessment of the participants' conceptions (especially when followed up with interviews). Unfortunately, due to their open-endedness, the VNOS questionnaires are not easy to administer, in particular when surveying large samples. Lederman et al. reported on a large study in which written and interview responses of participants were compared. From this study's results Lederman et al. infered the questionnaires' validity. Lederman et al. emphasized the necessity to establish intercoder reliability when analyzing participants' responses.

One instrument not included in Lederman's review (2007) is the Views About Science Survey (VASS; Halloun & Hestenes, 1996, 1998). The VASS includes three so-called 'scientific dimensions' – structure, methodology, and validity, and three so-called 'cognitive dimensions' – learnability, reflective thinking, and personal relevance (Halloun & Hestenes, 1998). Halloun (2001) reported a thorough approach of establishing validity by investigating several forms of validity. For example, item valitidy was explored by expert ratings concerning the content being assessed and concerning the expected expert answer as well as by a comparison of written and oral student responses. Based on these efforts' results, Halloun concluded that the VASS instrument was valid; unfortunately, he did not provide any statistical values like interrater reliability. Speaking in terms of Lederman's review (2007), the VASS would not meet content validity since its cognitive dimensions go beyond the scope of nature of science; and thus the VASS would not be a valid nature of science instrument. With respect to internal consistency, Halloun (2001) named several reasons for why classical reliability measures are not appropriate for the VASS. Instead, he argued that internal consistency may be assumed since "student average scores are about the same on all six dimensions" (p. 13). With respect to test-retest-reliability, Halloun reported about exit interviews two days after administering of the questionnaire and that "[v]irtually all these students reiterated the same answers they had indicated previously" (ibid.); again, no statistical values were provided. Different forms of the VASS being administered in different semesters were used to show the stability of the VASS measures, yet again statistical values were missing (cf. Halloun, 2001). Priemer (2003) translated the VASS into German and employed it on a German sample. This German version of VASS focused on physics only; whereas the English original version also covered chemistry and biology. Concerning validity and reliability, Priemer unfortunately did not provide any statistics gained from his studies, instead referring to what Halloun (2001) stated about the original instrument.

Recently, an instrument, for which psychometric quality criteria have been reported, has been introduced by Urhahne et al. (2008). These authors developed a five-option Likert-type instrument in German, the so-called 'Seven Scales of Nature of Science questionnaire' (SNOS). The authors reported a satisfying reliability for the questionnaire as a whole ($\alpha = .84$) as well as for each of the scales ($.52 < \alpha < .71$). The scales were elaborated from current research on Nature of Science and on epistemological beliefs. Consequently, there are some scales rather closely related to epistemological beliefs, and that therefore go beyond the scope of nature of science.

Conceptions of the nature of science versus epistemological beliefs

Instruments mixing up the nature of science and epistemological beliefs indicate that these two concepts are in close proximity to each other. However, nature of science is central to science education, whereas epistemological beliefs play a central role in psychological research in particular (e.g., Hofer & Pintrich, 1997; Schommer, 1990). Hofer and Pintrich (1997) provided a thorough and detailed review on relevant research concerning epistemological models of the past decades. This review makes clear that there are many different approaches to outline epistemological beliefs, but that the models do show similarities to some extent. For instance, most research on epistemological beliefs assumes that such beliefs can range on continuums from 'naïve' to 'sophisticated', whereby naïve beliefs are subject to developmental changes in childhood and adolescence. In their review, Hofer and Pintrich identify epistemological beliefs to be made up of two dimensions: a) the nature of knowledge, and b) the nature of knowing. Within the dimension of 'nature of knowledge' two factors were elaborated from the reviewed models: the certainty of knowledge and the simplicity of knowledge. Concerning the 'certainty of knowledge', a naïve belief would be that "absolute truth exists with certainty" (Hofer & Pintrich, 1997, p. 120) in contrast to knowledge being subject to change. Likewise, concerning the 'simplicity of knowledge', a naïve view is that "knowledge is [...] an accumulation of facts" (p. 120) rather than "highly interrelated concepts" (p. 120). The second dimension 'nature of knowing' is constituted by the factors 'sources of knowledge' and 'justification for knowing'. The naïve view concerning sources of knowledge involves that "knowledge originates outside the self and resides in external authority" (p. 120). The 'justification for knowing' refers to "individuals' justification of their claims" (Schommer, 1994, p. 296). On the naïve level, individuals do not allow for other opinions, instead thinking that knowledge is absolute.

According to Hofer and Pintrich (1997), these four factors (certainty, simplicity, sources of knowledge, and justification for knowing) make up the defining core of epistemological beliefs[5]. Comparing the concept of nature of science with these factors reveals three categories: (1) factors of epistemological beliefs, which have both an equivalent and an overlap, to aspects of the nature of science, (2) factors of epistemological be-

---

[5] Sometimes scales on the nature of learning are added (e.g., Schommer, 1990); however, Hofer and Pintrich (1997) did not include them in their model of epistemological beliefs as most of their factor analyses were not able to reveal Schommer's fifth factor.

liefs, which have no equivalent aspect of the nature of science, and (3) aspects of the nature of science, which have no equivalent factor of epistemological beliefs.

NOS (in the notion of Lederman, 2006) includes *Tentativeness* of scientific knowledge, which fits the 'certainty' aspect of epistemological beliefs quite well. Likewise, subjective influences on scientific knowledge (*Subjectivity* according to Lederman, 2006) to some extent relate to the factor of 'justification for knowing'. However, 'simplicity of knowledge', which is related to the structure of the body of knowledge, and the 'sources of knowing', which distinguishes between authority-given and self-constructed knowledge, have no equivalent aspect concerning any of the above discussed conceptions of nature of science (category 2). There are also aspects concerning nature of science that do not match any of the four factors of epistemological beliefs (category 3). For example, the *Role and relationship of scientific theories and laws*, and the *Embeddedness in culture* do not correspond to any of the four factors.

In summary, these three categories show that there is an overlap between epistemological beliefs and the concept of the nature of science, yet no complete identity between the two constructs. This partial overlap (category 1) is no surprise since Nature of Scientific Knowledge (NOS) itself refers to the *epistemology* of scientific knowledge. However, categories 2 and 3 show that this overlap does not mean one construct is included in the other (e.g., nature of science is not a subset of epistemological beliefs or vice versa). Thus, this consideration reveals that while they show parallels and overlap, epistemological beliefs and nature of science are still two individual constructs.

Students' conceptions of the nature of science

Despite the possible conflation of epistemological beliefs with nature of science and despite corresponding validity issues of respective instruments, research on the nature of science has revealed surprisingly consistent insights about students and teachers holding inadequate views about NOS[6] (Lederman, 2007). There have been several studies on students' conceptions about NOS (cf. Höttecke, 2001a; Lederman, 2007). In Germany, Meyling (1990, 1997) was one of the first to investigate students' conceptions of nature of science, even if he himself does not use the term 'nature of science'. Meyling employed open-ended as well as multiple-choice questionnaires, interviews and protocols of physics lessons to identify students' conceptions about the epistemology of science. Amongst other findings, Meyling (1997) revealed an insufficient view on, for instance, laws and theories. Some years later, Höttecke (2001a) provided a review of German and English research reports concerning conceptions about the nature of science, which also included Meyling's work. The most recent and comprehensive review of the past decades' studies that have been published in English was provided by Lederman (2007).

---

[6] Since this dissertation focuses on students, the teachers' conception will not be discussed.

Both authors, Höttecke and Lederman, point out that, in summary, students hold insufficient and inadequate conceptions concerning the nature of science. These inadequate conceptions are summarized in Appendix A. It is not claimed that the conceptions listed in Table A 1 are the most frequently found ones. The list may instead be seen as an overview of possible, inadequate conceptions.

Students' conceptions that were identified by Lederman (2007) more or less relate to the characteristics of scientific knowledge. Höttecke (2001a) classified students' conceptions in four themes: (1) The scientist as a person, the work of scientists, the conditions of scientific work; (2) scientific knowledge; (3) experiments as a part of instruction[7] and of research; (4) conditions of producing scientific knowledge. Comparing Lederman's and Höttecke's review shows a huge overlap. Most conceptions identified by Lederman can be found in Höttecke's themes (2), (3) and (4). No parallel could be found for theme (1). This is unsurprising because this theme is not included in the common aspects of nature of science (cf. Section 3.1). The only aspect that corresponds to this theme is the role of the scientific community. Themes (3) and (4) show some overlap to the Nature of Scientific Inquiry (cf. Table 1) when concerning the diversity of scientific procedures. Höttecke (2001a) and Lederman (2007) have both identified similar, inadequate conceptions of students concerning scientific knowledge.

Most of the inadequate conceptions identified by Höttecke (2001a) and Lederman (2007) cumulate in a realistic, empiristic, and/or ontological view of scientific knowledge. The influence of creativity, imagination and subjective factors on scientific knowledge is underestimated. Rather, science is viewed as reading off the truth (i.e., hard facts) from nature. From this point of view, hypotheses are an inferior type of knowledge: Only by a highly accurate and/or often repeated measuring process can true results be found, allowing hypotheses to mature into laws. Laws, on the other hand, are seen as fixed and certain knowledge because they represent the true nature. Likewise, the role of the scientific community is misjudged. What is valued as valid knowledge is seen as being independent from the social structure of the scientific community (e.g., research programs) and rather as being determined by only empirical evidence (because this is the truth). Finally, because it represents the truth found in nature, scientific knowledge is not seen as tentative but immune to changes.

Approaches to change conceptions of the nature of science

Based on the insight that students hold inadequate conceptions of nature of science, approaches were elaborated on how to teach such topics in school in order to develop adequate conceptions. Such approaches can be categorized into three types: enhancing

---

[7] Since this dissertation is not concerned with instructional practice, those conceptions that relate to experiments as a part of instruction are not taken into account.

understanding the nature of science by (1) scientific inquiry processes; (2) explicit activities; and (3) historical case studies.

Carey, Evans, Honda, Jay, and Unger (1989) developed a unit for 7th graders, which is categorized as the first type. The underlying idea of this unit was to convey process skills concerning scientific investigations "in a wider context of metaconceptual points about the nature of scientific knowledge" (Carey et al., 1989, p. 517). A central part of this unit was to reflect on the process of active construction of scientific knowledge. In Germany, Sodian, Thoermer, Kircher, Grygier, and Günther (2002; in more detail Grygier, 2008) adapted this unit and demonstrated that nature of science topics can be discussed even with young children. Likewise, Kishfe and Abd-El-Khalick (2002) investigated inquiry-oriented instruction and its influence on NOS views. The main finding of this study was that – in the scientific-inquiry oriented type of instruction – an explicit and reflective approach is more effective than an implicit approach. In other words, this means that inquiry-oriented instruction can enhance an adequate understanding about the nature of science, if nature of science aspects are explicitly reflected.

Explicit activities – the second type of approach – are the NOS activities proposed by Lederman and Abd-El-Khalick (1998), for instance. These activities are developed to address crucial NOS aspects and do not directly relate to scientific content. One of these activities used, for example, the so-called 'mystery tube', which is a system of ropes within a non-transparent tube. Pulling one rope causes another rope or several ropes to be pulled in the tube. When students are trying to explain what the inside of the tube looks like (based on their observations), the differentiation between observations and inferences can be discussed and illustrated. Likewise, the influence of imagination and creativity on knowledge can be demonstrated.

The third type of approach employs cases from science history. This historical approach is widely used to implement nature of science aspects in science education (e.g, Galili & Hazan, 2001; Höttecke, 2001b, 2004a; Kipnis, 1998; Klopfer & Cooley, 1963; McComas, 2008; Solomon, Duveen, Scot, & McCarthy, 1992). The European Union has recently provided funding for the research project 'History and Philosophy in Science Teaching (HIPST)' which aims to draw a connection between history of science and nature of science aspects in instruction (Höttecke & Rieß, 2009). A major advantage of the historical approach is that historical cases can be interwoven with science content "without taking additional classroom time" (Kipnis, 1998, p. 191). However, Abd-El-Khalick and Lederman (2000) provided tentative evidence that history of science courses promoted a change towards an adequate view concerning nature of science only if such aspects were addressed explicitly (see also Lederman, 2007).

Because the concept of nature of science is neither explicitly included in German standards nor in curricula, research on teaching and learning about the nature of science in Germany is still in its infancy (compared to other countries, like the United States). As illustrated above, Meyling (1990), Höttecke (2001b) and Priemer (2003) were concerned with research on students' misconceptions about the nature of science; Sodian et al. (2002) and Höttecke and Rieß (2009) provided approaches to teach aspects of the nature of science. By demonstrating several links between aspects of epistemology and current school physics education content, Leisen (2009) has recently encouraged German physics teachers to not shy away from including nature of science topics in their instruction. However, Leisen's appeal gives the illusion of the nature of science being easily taught by simply teaching physics as before. This is in clear contrast to the demand of an explicit and intense discussion of nature of science aspects (Lederman, 2007). In the end, the best teaching approach can only be determined through precise intervention studies.

4 Competence Regarding NOSI and NOS

In Chapter 1 the necessity to adequately understand the nature of science was emphasized with respect to socioscientific problems and discussions. Accordingly, a normative rationale was given on why an adequate understanding of the nature of science should be included in a contemporary notion of *Allgemeinbildung*, and thus, should be a goal of non-vocational education. Against the background of educational standards, competences were shown to be the current approach for setting educational goals (Chapter 2). Regarding the topic of nature of science, this means that mere knowledge about the characteristics of science is not sufficient for successfully and responsibly dealing with science-related social problems; rather, students need to achieve competence in the field of nature of science. Finally, Nature of Scientific Inquiry (NOSI) and Nature of Scientific Knowledge (NOS) were demonstrated to be two central, mostly agreed on components of the broad topic of nature of science (Section 3.1). Founded on these three aspects – understanding the nature of science as an educational goal for everyone; competence concerning the nature of science as an educational outcome; and NOSI and NOS as central components of the nature of science – research on competence regarding NOSI and NOS is necessary. Consequently, based on the definition of Weinert (2001) in the constraint form suggested by Klieme and Leutner (2006), competence regarding NOSI and NOS is defined as follows:

Competence regarding nature of science is understood as cognitive abilities and skills with respect to Nature of Scientific Inquiry (NOSI) and Nature of Scientific Knowledge (NOS) that is to identify and to reflect characteristics of scientific inquiry processes and of scientific knowledge. Accordingly, competence regarding

NOSI and NOS covers a range from simple cognitive activities up to higher order thinking skills[8].

Competence models create a foundation for the measurement and development of competences. A competence model typically differentiates components and levels of competence by thoroughly describing and delineating a given field and its levels of mastery (cf. Section 2.2). In the following section a competence model that specifies the above definition of NOSI and NOS competenece will be normatively outlined based on the research on competences and competence modeling as well as on the extensive research on nature of science presented in the preceding chapters. In the two subsequent sections, the components and levels of competence regarding NOSI and NOS are detailed.

4.1 The Competence Model

The model of competence regarding NOSI and NOS is based on the competence model that is currently used to benchmark German science education standards (ESNaS model, cf. Section 2.2). The ESNaS model comprises three dimensions: (1) *areas of competence*, (2) *cognitive processes*, and (3) levels of *complexity*. *Areas of competence* includes different content domains (e.g., the use of content knowledge, and the acquirement of knowledge) that refer to particular components of scientific competence. *Cognitive processes* corresponds to general strategies of solving a problem (i.e. task). Levels of *complexity* correspond to different types of information that have to be processed when solving an item. The ESNaS model has been chosen as a point of departure for two reasons. First, there is currently no model being empirically validated as a whole that operationalizes the German science education standards. The ESNaS model is based on studies that have shown that the idea of *complexity* is appropriate for defining competences (Kauertz, 2008). This finding holds true for several content areas (Bernholt, 2010; Kauertz, 2008); therefore, it can be reasonably assumed that *complexity* can be applied to further content areas to define competence levels. Moreover, first studies on the ESNaS model regarding *use of content knowledge* and *acquirement of knowledge* corroborated the expected effects of *complexity* and *cognitive processes* on item difficulty (Kauertz et al., 2010; Section 2.2). Second, using this model ensures the compatibility of the model developed within the framework of German science education standards. As demonstrated in Chapter 1, these standards do not explicitly include the concept of nature of science; however, an understanding of the nature of science was shown to be important to *Allgemeinbildung* (general education). Accordingly, the German science education standards should be extended by the concept of nature of science. To allow for a possibly use of this

---

[8] Since NOSI refers to only the *understanding about* scientific inquiry, competence regarding NOSI does not include inquiry-related performance skills as, for example, stating and testing an hypothesis.

project's findings with respect to extending the German standards by including the nature of science, the compatibility of the competence model to be developed with the German education standards is essential.

According to German science education standards, the area of competence termed 'acquirement of knowledge' covers the process of acquiring scientific knowledge through empirical investigations or through scientific modeling, for example. Understanding about the nature of science adds a meta-cognitive component to these cognitive competences. Understanding about the nature of science consequently fits into this area of competence. As discussed in greater detail in Section 3.1, NOSI and NOS (third and fourth column of Table 1) constitute two central components of nature of science. This is because: 1) aspects of both components were found within three major research reports on what nature of science content should be included in school science education; 2) both components are included (and differentiated in the same way) in the PISA framework of scientific literacy.

In accordance with the ESNaS model, the developed model of competence regarding NOSI and NOS embraces three dimensions (see Figure 3). The dimension *area of competence* is reduced to the concept of nature of science, in particular to the two components NOSI and NOS. To clearly signify that these two components do not constitute a complete area of competence, they are simply referred to as *content*. The dimension of *complexity* determines which type of information needs to be processed. The dimension of *cognitive processes* determines how such information needs to be processed. Both dimensions were adopted from the ESNaS model.

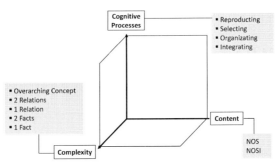

Figure 3. Competence model regarding Nature of Scientific Inquiry (NOSI) and Nature of Scientific Knowledge (NOS).

4.2 Core Aspects

To operationalize NOSI and NOS, aspects of Nature of Scientific Inquiry and Nature of Scientific Knowledge (see Table 1) that were suggested to be included in school science education by Lederman (2007; personal communication, 15.10.2007) were used.

According to Lederman, NOSI comprises eight aspects (e.g., *Inquiry procedures are guided by the question asked*; *Scientific data are not the same as scientific evidence*) and NOS encompasses seven aspects (e.g., *Subjectivity, Distinction between theory and law*). These aspects are condensed into three core aspects for each component, NOSI and NOS[9]. NOSI includes the core aspects: Scientific investigations (1) begin with a question; (2) embrace multiple methods and approaches; and (3) allow multiple interpretations. The core aspects of NOS are: Scientific knowledge is (1) subjective; (2) empirically based and inferential; and (3) tentative. Table 2 shows how these core aspects correspond to the aspects of NOSI and NOS as described in Section 3.1. In the following, the core aspects of NOSI and NOS are described in greater detail[10].

Scientific investigations begin with a question

Questions are a point of departure in all scientific investigations. A question can be used in different ways: It might ask for qualitative or quantitative criteria; it might focus on the generation of hypotheses or on testing a hypothesis. An important characteristic of a scientific question is that the products it asks for are phenomena of animate or inanimate nature. Moreover, it must be possible to process the question empirically, in principle. Even if the instruments have not been developed by the time of asking, the question might still be considered scientific if the respective instruments can be thought of in principle.

Scientific investigations encompass multiple methods and approaches

There is no exclusive, single scientific method. In science, there is no recipe that generates true knowledge automatically. Depending on various factors (the research question, available instruments, personal preferences, etc.) various methods can and must be applied (e.g., descriptive, correlative, experimental methods). Moreover, in science variables are not always systematically varied – this is only true when an experiment is conducted. Further, experiments are not the only method of scientific inquiry: There are observations, measuring, comparing, theoretical development, modeling, or simulation studies. No matter the method, they are being guided by the question asked: The method has to be chosen in such a way that its use makes answering the question possible (or that the proposed hypothesis can be tested). Moreover, a set of data is the output of a scientific method regardless of the method applied.

---

[9] The condensation as well as the description of core concepts has been elaborated in cooperation with Gary M. Holliday, Judith S. Lederman, and Norman G. Lederman.
[10] In similar wording, this description has also been included in Köller, Katzenbach, Mayer, Hartmann, Kremer, Wellnitz et al. (2008, p. 33-37).

Table 2. Condensation of aspects of of Nature of Scientific Inquiry (NOSI) and Nature of Scientific Knowledge (NOS) into core aspects.

Nature of Scientific Inquiry	
Core aspect: Scientific investigations…	Aspects
(1) begin with a question	Scientific investigations all begin with a question, but do not necessarily test a hypothesis.
	Inquiry procedures are guided by the question asked.
(2) embrace multiple methods and approaches	There is no single set and sequence of steps followed in all scientific investigations (i.e., there is no single scientific method).
	Inquiry procedures can influence the results.
(3) allow multiple interpretations	All scientists performing the same procedures may not get the same results.
	Scientific data are not the same as scientific evidence.
	Research conclusions must be consistent with the data collected.
	Explanations are developed from a combination of collected data and what is already known.
Nature of Scientific Knowledge	
Core aspect: Scientific knowledge is…	Aspects
(1) influenced by subjectivity	Subjectivity (Theory-ladenness)
	Creativity
	[Cultural embeddedness][11]
(2) empirically based and inferential	Empirically based
	Distinction between observation and inference
	Distinction between theory and law
(3) tentative	Tentativeness

Scientific investigations allow multiple interpretations

Different scientists can interpret a set of data differently. Various conclusions can be drawn depending on various factors (the research question, a scientists' experience and imagination, etc.). Thus, there is not a single, exclusive correct interpretation of a set of data. Therefore, all conclusions have to be criticizable, in principle. Moreover, interpretations are made based on empirical data. Data are interpreted in relation to knowledge already known, to currently accepted theories and models. The interpretations must be developed from the known body of knowledge and must be logically consistent with the data. This is done according to currently accepted rules of evidence.

---

[11] Cultural embeddedness relates to the core aspect as indicated. However, in the presented project, this aspect has not been included, since previous work showed that this aspect is very difficult to be understood by students (N. G. Lederman, personal communication, 15.10.2007).

Scientific knowledge is influenced by subjectivity

Scientific knowledge does not come out of nowhere – it is generated by scientists, by human beings. These people are influenced by individual and social factors. Therefore, the process of scientific inquiry and scientific knowledge is not strictly objective but is also influenced by these factors. Individual and social factors influence which aspects scientists focus their research on and how they deal with it; e.g., the choice of the research question depends on a scientist's interest and private socialization. The process of scientific inquiry can be influenced by scientists' creativity, financial backgrounds, their roles during the process, along with any influence their current worldviews may have and so forth. Social factors refer to current beliefs and views of community and scientific community. Currently relevant fields of research and research questions, along with currently accepted methods, models, and theories are all influencing factors. Yet, being influenced by subjective factors does not imply that scientific knowledge is arbitrary.

Scientific knowledge is empirically based and inferential

Not every format of knowledge or way of generating knowledge is said to be scientific. A characteristic of science is that it is based on empirical investigations and their interpretations. Basic criterion of scientific knowledge is that it is generally bound to empirical reasoning. Scientific production of knowledge consists of obtaining data among natural systems by methods like observing, classifying, experimenting or simulating. These data are, however, not the same as scientific knowledge. Scientific knowledge is inferred from the interpretation of these data, which is guided by theories. Knowledge developed in this way is represented through theories, models, laws, and natural constants. Laws identify relationships among observable phenomena, in the way that Ohm's law, for example, describes the relation between resistance, voltage, and current in an electric circuit under certain conditions. Theories are inferred explanations for natural phenomena. Theoretical models for these explanations can be conceptual, mathematical, or physical.

Scientific knowledge is tentative

Scientific knowledge is never absolute or certain. This tentativeness is, among other things, caused by the above-mentioned aspects of subjectivity and their interdependency. Modifications of scientific knowledge can occur with reference to theory or method. With reference to theory, a modification is manifested by the reinterpretation of data. On the one hand, the reinterpretation can be influenced by a newly developed or a reworked theory. On the other hand, data can be reinterpreted from another perspective. This reinterpretation could be influenced by another theory that is part of another field of research. These reinterpretations can lead to new knowledge. A modification with respect to method is manifested by the production of new data. When new methods are developed, new data can be recorded that could not be accessed before. Improving methods can also lead to new knowledge, as new methods may reveal a finer resolution or a better signal to noise ratio and thus allow for new insights, for example.

4.3 Complexity and Cognitive Processes

The core aspects of NOSI and NOS serve to specify the *content* a particular competence refers to. Contrastingly, *complexity* and *cognitive processes* serve to distinguish between different high and low competences. *Complexity* levels define the ability to handle different types of information. *Cognitive processes* refer to different strategies of information processing.

Complexity

The idea of complexity is based on the principle of elements of knowledge and their combination (Kauertz et al., 2010; Neumann et al., 2007). The lowest complexity refers to the handling of single elements of knowledge. The highest complexity corresponds to complex knowledge structures consisting of a network of several elements and relations between those elements. Here, the smallest units are called 'facts'. In the context of NOSI and NOS, facts are defined as terms, attributes, simple descriptions of situations, or simple statements. Interrelations between those facts represent the next more complex type of information ('relations'). Concerning NOSI and NOSI, such interrelations can be, for example, of causal, conditional or relational nature. The most complex type of information is represented by so-called 'overarching concepts'. Overarching concepts are highly generalized; they do not depend on a specific situation and, thus, can be applied to several situations or problems. Within the competence model regarding NOSI and NOS, overarching concepts are operationalized by the six core aspects of NOSI and NOS, which were illustrated in Section 4.2. Processing facts is assumed to be easier than processing relations, which again is assumed to be easier than processing an overarching concept. Likewise, an increase in the amount of elements of information is assumed to generate difficulty. This means, for example, that processing one fact is assumed to be easier than processing two facts. In summary, the following five levels are assumed to be ordered hierarchically and to make up the *complexity* dimension:

Level I: One Fact

Level II: Two Facts

Level III: One Relation

Level IV: Two Relations

Level V: Overarching Concept

Cognitive processes

To solve problems (e.g., as they are given in test items) particular cognitive procedures have to be employed. According to Kauertz et al. (2010), these processes are characterized by three criteria: the relation between given and expected information; the necessity to build connections; and the similarity between a given situation and an answer situation. If expected information is identical to given information, students just have to replicate information in order to solve an item. If expected information is a subset of given information, students have to decide on the needed information. If expected information is an extension of given information, students have to involve other information to solve an item. Along these three cases, cognitive capacity is increasing. Likewise, the necessity to build connections increases the cognitive requirements of the cognitive process. Finally, a cognitive process is assumed to require more cognitive capacity if the similarity between the given and expected situations decreases. Table 3 displays how four cognitive processes evolve from different combinations of these three criteria. Accordingly, the *cognitive processes* dimension is made up by the following hierarchy:

(1)   Reproducing
(2)   Selecting
(3)   Organizing
(4)   Integrating

Table 3. Definition of *cognitive processes* according to Kauertz et al. (2010; translation by the author).

		Criteria		
		Relation between given and expected information	Necessary to build connections	Similarity between given and expected situation
Cognitive Processes	Reproducing	identical	no	high
	Selecting	subset	no	high
	Organizing	extension	yes	high
	Integrating	extension	yes	low

Complexities and cognitive processes

The combination of one core aspect of NOSI or NOS, one complexity level and one cognitive process defines a particular sub-competence. Combining the dimensions of *complexity* and *cognitive processes* reveals a 5x4 matrix corresponding to 20 conceivable compositions (Figure 4). However, not all of these 20 sub-competences are meaningful: Organizing one fact is not reasonable, because organizing as a cognitive process relates to several elements being organized following an order; in other words, organizing requires at least two facts. Likewise, integrating one or two facts is not reasonable because facts are bound to a concrete situation. For example, the fact that 'Chadwick interpreted data' is bound to a particular situation and cannot be applied to another situation, such as Newton's work on light. Facts being bound to a concrete situation are in opposition to integrating, which corresponds to the abstracting or the transferring of knowledge. In summary, 17 compositions of *complexity* level and *cognitive processes* are meaningful to be taken into account. Referring to NOSI, these 17 compositions are depicted white cells in Figure 4. For clarity reasons, the matrix for NOS is omitted in this figure, yet it is identical to the one for NOSI.

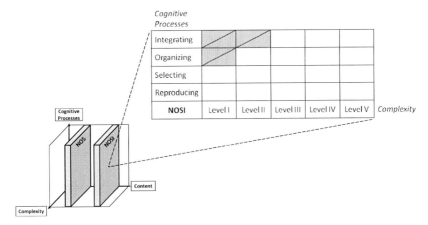

Figure 4. How item matrix of *complexity* and *cognitive processes* relates to the competence model regarding NOSI and NOS.

Research Question and Hypotheses

The aim of this dissertation is to develop and validate a model of competence regarding Nature of Scientific Inquiry (NOSI) and Nature of Scientific Knowledge (NOS). Thus, the primary research question focuses on the model's validity:

To what extent does the model of competence regarding NOSI and NOS show empirical validity?

Demonstrating validity requires the integration of various sources of validity (cf. Kane, 2006; American Educational Research Association [AERA], American Psychological Association [APA], & National Council on Measurement in Education [NCME], 2004). This study highlights aspects of construct and criterion validity. In this context, construct validity relates to whether the structure of the model hypothesized can be found from empirical data (Chapter 0). As for criterion validity, (1) discriminant validity and (2) a variation of curricular validity[12] are examined (Chapter 6). Discriminant validity focuses on distinguishing between competence regarding NOSI and NOS on the one hand and other competence-related variables on the other. Curricular validity corresponds to comparing two different samples, which are assumed to differ as a result of a differing emphasis that is put on the nature of science in respective standard documents.

5 Hypotheses Related to Construct Validity

The construct validity of the competence model devised in Chapter 4 refers to its internal structure. In particular, the assumed structure of the three dimensions *complexity*, *cognitive processes* and *content* needs to be demonstrated to correspond to empirical data.

---

[12] This type of validity is referred to as 'curricular validity' since the performance of two samples is predicted based differences in standard documents. This is not the tradidtional meaning of curricular validity; however, for reasons of clarity, the term 'curricular validity' is used here.

The dimension of *complexity* describes the type of information to be processed by a student (see Section 4.3). The following five *complexity* levels are assumed to make up this dimension: one fact (Level I), two facts (Level II), one relation (Level III), two relations (Level IV), and overarching concept (Level V). Additionally, these levels are assumed to be ordered hierarchically and to influence an item's difficulty. In order to investigate these assumptions, one first has to ensure that *complexity* represents one single construct – in contrast to each level representing one construct (H1a). Then, *complexity*'s influence on item difficulty (H1b) and its order has to be examined (H1c). Accordingly, the first hypothesis (H1) consists of three sub-hypotheses:

(H1) The assumed structure of the dimension *complexity* is represented by the data.

  (H1a)   *Complexity* represents a unidimensional variable.

  (H1b)   *Complexity* shows a statistically significant effect on item difficulty.

  (H1c)   Mean item difficulty μ per *complexity* level increases with increasing complexity level: $\mu_I < \mu_{II} < \mu_{III} < \mu_{IV} < \mu_V$.

Cognitive processes make up the second dimension of the competence model regarding NOSI and NOS (see Section 4.3). This dimension describes how information has to be processed. It is made up of the following four processes: reproducing (rep), selecting (sel), organizing (org), and integrating (int). Like *complexity*, *cognitive processes* are assumed to be ordered hierarchically and to influence an item's difficulty. Again, *cognitive processes* have to be investigated concerning the number of constructs they represent (H2a) before their influence on item difficulty (H2b) as well as their order (H2c) can be scrutinized. In summary, the following hypotheses regarding *cognitive processes* are studied:

(H2) The assumed structure of the dimension *cognitive processes* is represented by the data.

  (H2a)   *Cognitive processes* represent a unidimensional variable.

  (H2b)   *Cognitive processes* show a statistically significant effect on item difficulty.

  (H2c)   Mean item difficulty μ per *cognitive process* increases with increasing *cognitive process* level: $\mu_{rep} < \mu_{sel} < \mu_{org} < \mu_{int}$.

The remaining dimension of the competence model is the *content* dimension that consists of two components of the nature of science: Nature of Scientific Inquiry and Nature of Scientific Knowledge. In line with current research on nature of science (cf. Section 4.2) NOSI and NOS are viewed as separate concepts. Thus, competence regarding NOSI and competence regarding NOS are reasonably assumed to be separate from each other (H3a). In contrast to the two dimensions discussed before, aspects of NOSI and NOS are not assumed to be ordered hierarchically and, thus, are not assumed to influence item difficulty (H3b). In short, the following hypotheses are investigated to explore the structure of the *content* dimension:

(H3)   Competence regarding NOSI can be distinguished from competence regarding NOS.

    (H3a)   NOSI and NOS items correspond to two separate traits.

    (H3b)   Aspects of NOSI and NOS do not show a significant effect on item difficulty.

Each of the above three sets of hypotheses (H1-H3) deals with a component of the competence model regarding NOSI and NOS. All of them elucidate a particular part of the internal structure of the model. Altogether, they contribute to the model's construct validity.

6   Hypotheses Related to Criterion Validity

In contrast to construct validity, criterion validity is scrutinized from an external point of view. Two types of criterion validity are focused on here: discriminant validity and curricular validity.

Investigating discriminant validity, especially, involves other traits being contrasted to competence regarding NOSI and NOS. As discussed in Section 2.1, competence refers to cognitive abilities. However, these cognitive abilities can be learned and are domain-specific. Thus, competence should be distinguished from the kinds of cognitive abilities bound to natural development, like intelligence (H4a). Moreover, competence has to be distinguished from general, domain-unspecific abilities, like reading abilities (H4b). Even if *cognitive processes* that require only low cognitive capacity (i.e. reproducing and selecting) are quite close to general reading abilities, competence concerning NOSI and NOS is assumed to involve more than just the ability to read and understand a text. Additionally, affective and attitudinal aspects should be investigated and differentiated from competence on NOSI and NOS (H4c). As illustrated in Section 2.1, competence embraces cognitive and motivational aspects that should be modeled and assessed separately in order to elucidate their interrelationship. Finally, the relation between competence on NOSI and NOS on the one hand and views of the nature of science and science-related epistemological beliefs on the other hand has to be explored. A differentiation between those constructs is to be expected (H4d) because general knowledge – and thus competence referring to applied knowledge - and views (or beliefs) represent separate traits (cf. Mayer, 2008). However, an adequate view of the nature of science and sophisticated science-related epistemological beliefs probably contribute to competence regarding NOSI and NOS. This is why this last relation is expected to be stronger than the others (H4e). In summary, the following hypotheses contribute to the exploration of discriminant validity:

(H4) Competence regarding NOSI, and NOS, respectively, can be differentiated from other competence-related abilities and traits.

  (H4a)   Competence regarding NOSI and NOS can be distinguished from general cognitive abilities.

  (H4b)   Competence regarding NOSI and NOS can be distinguished from reading abilities.

  (H4c)   Competence regarding NOSI and NOS can be distinguished from interest and self-belief concerning physics and physics education.

  (H4d)   Competence regarding NOSI and NOS can be distinguished from views on the nature of science and science-related epistemological beliefs.

  (H4e)   Competence regarding NOSI and NOS is more strongly related to views on NOS than it is related to general cognitive abilities, reading abilities, and affective and attitudinal traits regarding physics and physics education.

The second aspect of criterion validity that is explored in this study refers to curricular validity. With respect to this facet, students from Germany will be compared to students from the United States. There is a long tradition of NOSI and NOS in science education research in the American academic community. Moreover, aspects of NOSI and NOS are explicitly mentioned in U.S. education standard documents. In contrast, German standards and curricula contain such aspects only implicitly; accordingly most German teaching is assumed to cover the topic of nature of science only implicitly. Therefore, it is assumed that U.S. students are exposed to more intense and explicit NOSI and NOS instruction than German students are. As a result, U.S. students are reasonably assumed to outperform German students (H5):

(H5) Concerning NOSI and NOS competence, U.S. students show higher performance than German students.

Altogether, the five main hypotheses H1-H5 cover a wide range of validity aspects, of construct and criterion validity, in particular. Figure 5 illustrates how the different hypotheses contribute to establishing validity of the competence model. Hypotheses H1-H4 are each indicated to be constituted by subhypotheses that have been detailed above. The subhypotheses are investigated to examine the main hypotheses.

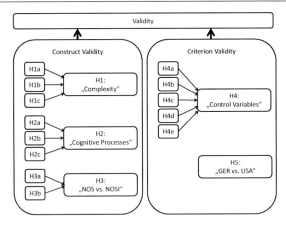

Figure 5. Framework for the hypotheses (H) investigated in this study.

Methods

To investigate the validity of the developed competence model (Chapter 4), the model needs to be operationalized into test items. Therein, operationalization means that, based on a particular student's response to an item, inferences can be made on the student's ability to apply a particular *cognitive process* on a particular *complexity* level with respect to a particular core aspect of NOSI or NOS. Therefore, items that represent all possible sub-competences have to be developed; that is, all possible combinations of elements from the model's three dimensions have to be represented by items. Systematic development ensures that the test items fit the model, and thus is mandatory for inferring conclusions regarding the competence model from the collected data. Items, that meet this proximity to the model, then need to be employed to gather empirical data on students' competence regarding NOSI and NOS. This way, aspects of construct validity can be explored (Chapter 0). In order to investigate the hypotheses dealing with particular aspects of criterion validity (Chapter 6), these items need to be administered together with other tests and questionnaires (discriminant validity) and in different samples (curricular validity).

The research design to explore the model's validity is detailed in Chapter 7. To analyze the obtained data, methods of classical test theory (CTT), probabilistic test theory (PTT), Rasch analysis in particular, as well as inferential statistics are used. The methodology used for data analysis is addressed in Chapter 8.

7   Research Design

The following research design was used to approach the question for validity of the developed competence model: First, the model was operationalized into test items. Second, these items as well as additional control instruments were employed to investigate construct and discriminant validity in the first study. Finally, the developed test items were employed in a sample including students from Germany and the USA to corroborate the Study 1 findings concerning criterion validity and to additionally investigate curricular validity. Accordingly, Section 7.1 details how the three dimensions of the model were realized in test items. Study 1, focusing on the investigation of criterion and discriminant validity, is described in Section 7.2. The following Section 7.3 addresses Study 2, which focuses of criterion and curricular validity.

7.1 Development of Test Items

The test items are the operationalization of the competence model regarding NOSI and NOS, which was described in Chapter 4. With respect to the three dimensions of the model, *content*, *complexity*, and *cognitive processes*, each item needs to represent a level of complexity, a cognitive process and a core aspect of NOSI or NOS. The following paragraphs detail how these three characteristics were realized during item development. To account for competence being domain-specific (Section 2.1), test items needed to be embedded in particular contexts. These contexts are also detailed here.

Content

The *content* dimension consists of the concepts of Nature of Scientific Inquiry and Nature of Scientific Knowledge (Section 4.2). These two components are each represented by three core aspects. Accordingly, items were developed to survey the understanding of these six core aspects:

- **NOSI**: Scientific investigations…

  (1) … **begin with a question**. Scientific investigations do not necessarily aim to test hypotheses and, thus, do not necessarily have an hypothesis as a starting point. Investigations are initiated and guided by research questions. Questions influence and determine approaches of data collection and interpretation.

  (2) … **encompass multiple methods and approaches**. Scientists conduct experiments but they also employ other approaches. For example, they make observations or they make use of theoretical approaches. There is not only one single scientific method.

  (3) … **allow multiple interpretations**. Depending on research questions and underlying theories, data can be interpreted differently. There is not one single correct interpretation and/or inference of a set of data. Based on the same set of data, different researchers can come to different results.

- **NOS**: Scientific knowledge is…
  - (4) … **influenced by subjectivity**. Scientific knowledge is produced by human beings. Therefore, it is not completely objective, but is influenced by the character of a researcher, for instance.
  - (5) … **empirically based and inferential**. The baseline of scientific knowledge is empirical data. However, data by itself are not knowledge; they have to be interpreted in order to obtain meaning.
  - (6) … **tentative**. There is no absolute scientific knowledge. Scientific knowledge can change, e.g., due to the development of new instruments or due to a reinterpretation of data based on a new theory.

Complexity

The dimension *complexity* consists of five hierarchically ordered levels that correspond to differing complexities of information (Section 4.3). During item development, these levels were realized as follows:

  - (I) **One Fact**: Answering an item requires either a simple description of a situation or an ascription of a term, attribute, etc.
  - (II) **Two Facts**: Answering an item requires two simple descriptions of a situation, a simple description of two situations, or an ascription of two terms, attributes, etc.
  - (III) **One Relation**: Answering an item requires an explicit interrelation (e.g., causality, condition, relation).
  - (IV) **Two Relations**: Answering an item requires two explicit interrelations (e.g., causalities, conditions, relations).
  - (V) **Overarching Concept**: Answering an item requires the understanding of one aspect of NOS or NOSI as a whole, or a central part of such an aspect to be applied.

Cognitive processes

The dimension of *cognitive processes* differentiates between four hierarchically ordered strategies of information processing (Section 4.3). In particular items, *cognitive processes* were realized as follows:

  - (5) **Reproducing** information. The correct answer is already explicitly given in the stem. To solve an item, this information has to be identically reproduced. Wrong answer options do not appear in the stem.
  - (6) **Selecting** information. The correct answer is already explicitly given in the stem. To solve an item, this information has to be selected from other information also provided in the stem. Item stems also contain wrong answer options.

(7) **Organizing** information. Several pieces of information are explicitly given in the stem. To solve an item, those pieces have to be ordered (e.g., a causal relation has to be established).

(8) **Integrating** information. Information has to be transferred to and integrated into a new situation. Information has to be decontextualized from the situation given in the stem by, e.g., generalizing.

Item matrix

Items were developed to represent <u>one</u> aspect of NOSI or NOS, <u>one</u> *complexity level*, and <u>one</u> *cognitive process*. Based on the definitions of *complexity* levels and *cognitive processes*, three combinations of *complexity level* and *cognitive process* – organizing and integrating one fact, and integrating two facts – were shown to be not reasonable (Section 4.3). During item development a fourth combination was excluded. Operationalizing the selection of an overarching concept would require at least two concepts to be provided explicitly in an item's stem. This would drastically increase stem length. To keep reading ability from becoming an unintended influence, items were developed to be as short and easy as possible. Therefore, the combination of 'selecting' and 'overarching concept (Level V)' was not included in item development, either. All in all, 16 combinations of *complexity level* and *cognitive process* were considered for item development. These are depicted as cells with a white background in Figure 6.

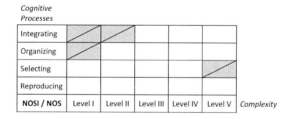

Figure 6. Combinations of *complexity* levels and *cognitive processes*. Grey cells were not included in item development.

Context

To account for competences' domain-specificity (Section 2.1), items were embedded in particular contexts. In Section 3.2, cases of science history were shown to be an appropriate and often used context to address the nature of science in school. Consequently, exemplary cases of physics history were chosen as contexts for the items. Such historical cases describe how particular physics knowledge was produced. For example, one applied context described the development of different models of the solar system – referring to Aristotle's, Ptolemy's, Copernicus's and Kepler's work. A historical case was seen to be suitable for item development if it illustrated core aspects of NOSI and NOS. In order to account for unintentional influences of these contexts, the cases were spread over several physics topics and periods (cf. Appendix B.1). For each historical case, between three and thirteen items were developed.

For item development, a particular historical case was summarized from physics history literature. Based on this summary, all necessary information related to the historical case was worded in the most comprehensible way possible. As a result, short 'stories' on physics history were produced and used as item stems. These 'stories' included all relevant background information concerning physics content and physics history. In doing so, items did not require students' competence to solve items on physics content knowledge, and as a consequence, surveying students' understanding of physics content or physics history was avoided. In line with the definition of NOSI/NOS competence (Chapter 4), students were then asked to identify and reflect on aspects of NOSI and NOS from these historical contexts.

Example items

The context 'The Universe' addressed different models of the universe, which were devised by Albert Einstein and Georges Lemaître, among others. Figure 7 shows two example items related to this context. The examples illustrate that all items include four options out of which only one is correct. Wrong options were developed based on the typical, inadequate conceptions about the nature of science (Table A 1). Moreover, the example items illustrate that some item stems provide abstract information about a particular core aspect of NOSI or NOS (here in italics), before the actual 'story' is given. The addition of this abstract information was necessary in order to operationalize particular combinations of *complexity* level and *cognitive process* (e.g., reproducing, selecting or organizing an overarching concept). The cognitive processes require all necessary information to be given explicitly in the stem, while the complexity level requires the (abstract) understanding of a core aspect.

Item 1 in Figure 7 focuses on the NOSI aspect of *Scientific investigations embrace multiple methods and approaches*: The historical case explicitly opposes a theoretical approach to an observational approach. To solve item 1, a student has to extract this circumstance on a general level. This addresses the aspect on the whole. Therefore, item 1 is assigned complexity level V (*overarching concept*). Additional information in the beginning already mentions this concept so it only has to be *reproduced* in order to solve the item.

The Universe

In science, there is not only one way to get results. Scientists conduct experiments and modify variables in order to collect data, but they also make observations, for example. Moreover, there are scientists that develop theories in order to explain observations and measurement results.

In the early 20th century, there were different scientific models describing the universe. Albert Einstein and Willem de Sitter thought of a universe that was static and rested in itself. Georges Lemaître, thought of a universe that was expanding and getting larger over time. From this model, Lemaître derived a law on the motion of neighboring galaxies. Independently, the physicist Edwin Hubble made observations and calculations from which he derived the same law. Combining both of their work resulted in a view on the origin and the development of the universe. Using it, Hubble's observations could be explained and it is still accepted, today.

Item 1 (gal7): Scientific investigations embrace multiple methods and approaches/ Overarching Concept/ Reproducing

What does Hubble's and Lemaître's work show about science in general?

- ☐   In science, there is only one method to follow in order to get results.
- ☐   Scientists have to work through experimental instructions to find the right results.
- ☐   In science, theories do not generate knowledge as precisely as experiments do.
- ☑   Scientists use different and various approaches in order to gain new knowledge.

Item 2 (gal8): Scientific knowledge is empirically based and inferential/ Two Facts/ Selecting

Which of the following statements best describes Hubble's research on the universe?

- ☐   Hubble investigated the universe and explained a theory.
- ☑   Hubble made observations and inferred a law.
- ☐   Hubble developed a law and it was proved correct.
- ☐   Hubble collected data and the data were a new law.

Figure 7. Two example items of the context 'The Universe'. Location in the item matrix is indicated. Correct options are checked.

Item 2 in Figure 7 is assigned complexity level II (*two facts*) and the cognitive process of *selecting*. To solve an item, a student has to process (1) that Hubble made observations (first fact), and (2) that he inferred a law (second fact). Both facts are given in the stem, but the investigation of the universe (first option) and a law's development (third option) are also mentioned in the stem albeit not in direct relation with Hubble. Thus, the student has to select the correct facts. The correct option addresses the NOS aspect of *Scientific knowledge is empirically based and inferential*. The statement that "Hubble made observations and inferred a law" already indicates that there is an interplay between empirical observation and inference.

7.2 Study 1

Study 1 aimed (1) to investigate if the hypothesized structure of the competence model is represented by empirical data (construct validity; hypotheses H1, H2, and H3), and (2) to explore the relationship between competence regarding NOSI and competence regarding NOS, as well as other competence-related constructs (criterion validity; Hypothesis H4). For this purpose, a 'NOSSI test' was compiled from the items developed. Additional tests and questionnaires were employed to survey students' cognitive abilities, reading abilities, beliefs about the nature of science, and their interest and self-belief. All of these tests were administered on a sample of German students.

Instruments

For Study 1, 107 NOSSI-test items were developed covering all possible cells of the model of competence (cf. Appendix B). In doing so, attention was paid (1) to having no vacant cell for NOSI and NOS, respectively, (2) to having marginal totals that allow further analyses, and (3) to having about the same amount of items related to NOSI as related to NOS. In Appendix B.2, Table B 2 and Table B 3 provide an overview of these distributions for both NOSI and NOS, as well as for each of the six core aspects. Because of a planned international comparison, all NOSSI items were developed in English and translated into German.

The items were compiled into booklets using a rotated multi-matrix sampling. Preparing for this, items were disjointedly grouped into twenty-two item sets to cover a processing time of approximately 10 minutes each. Each booklet contained three sets. Two sets overlapped between every two booklets. Each item set was included in at least three booklets. To avoid positioning effects, the sets' positions changed between these three booklets. When compiling the booklets, incompatibility of items was taken into account (i.e., no item should contain the answer to another item). All in all, items were distributed over 24 booklets (cf. Appendix B.5). Booklets contained between 13 and 16 items according to an estimated processing time of 30 minutes. Each booklet, then, represented a 'small' NOSSI test. Even if 107 items were included in the data gathering of Study 1, one item was excluded after data collection, since its historical soundness was put into question.

To survey students' cognitive abilities, one nonverbal scale (N2) of the cognitive ability test KFT (Heller & Perleth, 2000) was used. In this test, students were given a pair of figures and a single figure. Students then needed to decide which figure (out of a selection of five) together with the single figure would create a pair similar to the pair given. This test was limited to a processing time of 8 minutes. To avoid cheating, two different forms of KFT-N2 were used. In addition to this nonverbal scale, KFT usually embraces quantitative and verbal scales, too. However, since NOSSI items do not require calculation abilities and since reading abilities were surveyed differently, those scales were not included in this study.

The *Lesegeschwindigkeits- und -verständnistest* LGVT (reading speed and reading ability test; Schneider, Schlagmüller, & Ennemoser, 2007) was employed to survey students' reading abilities. The LGVT consisted of a text students were asked to read. This text contained gaps with a choice of three words. Students then had to decide which of those words would reasonably fill such a gap. After a processing time of 4 minutes, students had to indicate how far they had gotten. Reading comprehension was determined based on the number of correctly filled in gaps. Reading speed was determined based on the number of words read.

Beliefs about the nature of science were surveyed using Seven Scales of Nature of Science (SNOS, K. Kremer, personal communication, February 10, 2009; Urhahne, Kremer, & Mayer, 2008[13]). The SNOS questionnaire addresses the following seven scales of NOS: (1) Source (Q, 6 items), (2) Certainty (S, 8 items), (3) Development (E, 8 items), (4) Justification (R, 11 items), (5) Simplicity (K, 5 items), (6) Purpose (Z, 7 items), and (7) Creativity (V, 7 items). As illustrated in Section 3.2 these seven scales are a mixture of beliefs about the nature of science and science-related epistemological beliefs[14]. Each item consists of a short statement and five Likert-options of agreement. By random sampling, four booklet versions were compiled to prevent students from cheating. While being administered the test, students were informed that their beliefs and opinions were to be surveyed and that there would be no right or wrong answer to account for influences of social desirability. Students were given 20 minutes of processing time. It was ensured that all students had enough time to answer all items.

To assess students' interest and self-belief in physics, the Students' Interest and Self Concept questionnaire SIS (Kauertz, 2009) was modified. Originally, this questionnaire was developed for elementary and early secondary students, but was adapted for use in the sample of this study[15]. The so-called 'SIS$_{ad}$' consists of three scales regarding interest: situational interest in physics lessons (FI, 7 items), out-of-school interest in physics (AI, 6 items), and personal interest in physics (SI, 5 items). Moreover, it contains three self-belief scales: experienced competence in physics (EK, 4 items), self-efficacy regarding physics (SW, 5 items), and self-concept regarding physics (SK, 7 items). Items of FI, SI, EK, SW, and SK scales consist of a statement and four options of agreement. Out-of-school interest in physics (AI) was surveyed by asking how often students deal with physics topics outside of class. Thus, AI items contained four frequency related options. Again, to account for influences of social desirability, students were told that their opinion was to be surveyed and that there would be no right or wrong answer. Additionally, students were assured that their teachers would not be informed about their answers on this questionnaire. Students were given 10 minutes to work on the items.

---

13 The used version of this questionnaire is not the one discussed by Urhahne, Kremer, & Mayer, 2008; yet, it embraces the same scales, consists of more items, and represents the most recent version.
14 For readability reasons, the SNOS will be referred to as 'assessing beliefs about the nature of science'.
15 Changes were only related to the courses the questions referred to. While the original version referred to science lessons typical for German elementary school, the adapted version referred to physics lessons being appropriate for the investigated sample.

Altogether, students participated in a session of approximately 90 minutes. Instruments were used in the following sequence: (1) NOSSI test, (2) KFT, (3) LGVT, (4) SNOS, (5) SIS$_{ad}$. Tests that were more mentally challenging for students were administered first. Proceeding in this sequence was thought to account for fatigue effects.

Sample

NOSSI items were developed in accordance with the model of competence that is used in the evaluation of educational standards in Germany at the end of Grade 10 (cf. the ESNaS project, Section 2.2). Therefore, German 10th graders were chosen for this study. Altogether, 1086 students participated in Study 1. However, data from only $n = 1080$ students were included in data analysis. One student was excluded due to being an exchange student; another only spoke rudimentary German. Two students obviously did not thoroughly and seriously participate. Another two students had not finished with their NOSSI booklets after 30 minutes. Students who were included in data analysis were from *Hauptschule* (10.6 %), *Realschule* (33.8 %) and *Gymnasium* (55.6 %)[16] in North Rhine-Westphalia. On average, students were 16.1 years old (based on 1054 cases); 48.8 % of the participants were female (based on 1043 cases).

7.3 Study 2

Study 2 also explored internal and external validity aspects concerning the competence model for NOSI and NOS. Two aims were pursued in Study 2: Its first goal was to corroborate the findings of Study 1 concerning the competence model's internal structure (construct validity, hypotheses H1, H2, and H3). The second goal was to explore the curricular validity of the competence model (hypothesis H5), which involved the comparison of a German and a U.S. subsample with respect to their NOSI and NOS competences.

Instrument

In Study 2, all 106 NOSSI items that had been included in the data analysis of Study 1 were used. In principle, the items were the same. However, all items were revised before they were employed for a second time. To ensure item quality, the German NOSSI item stems were reviewed by an expert in physics history. This review process revealed that all contexts were proper, in principle. Yet, the presentation of the stories was made clearer, and the wording of all items was revised and simplified. After these revisions on the German version, the English version was adapted. During the adaptation process, the German item version was again revised in some cases in order to adjust the German and English versions. Such adjustment was meant to ensure that the German and the English item versions were as equivalent as possible. All revisions referred only to the wording of

---

16 For an overview and description of the different school types in Germany and its tracking system, see Döbert, 2007.

the items; the items' main ideas were kept. In particular, the categorization of *complexity level*, *cognitive process*, and core aspect of NOSI/NOS remained unchanged.

After this revision process, the items were compiled into booklets. Twelve booklets with an expected processing time of 40 minutes were compiled from the German items. This reassembling was due to the NOSSI test being the only instrument administered in Study 2, and due to the experiences of Study 1 in which the students took less time to solve the items than expected. Again, a multi-matrix sampling was used to distribute items over the booklets (cf. Appendix B.6): Items were compiled into 12 disjointed item sets, and each booklet contained two sets to guarantee an overlap between every two booklets. The U.S. booklets were thought to be of the same compilation as the German booklets. However, a pilot study with this booklet version revealed that the students did not work thoroughly (e.g., the teacher observed students arbitrarily checking or disconti- nuing their work in the middle of the test), probably because they were not used to such a large amount of items, became discouraged by the booklets and, thus, lost their motiva- tion. As a consequence, the U.S. booklets were shortened: In the main study, the U.S. booklets contained only one item set; therefore U.S. booklets were not overlapping. Like the German students, the U.S. students were given 40 minutes to work on the items.

Sample

For the second study, German and U.S. students were surveyed. To have equivalent age groups (about 16 to 17 years old), 10th graders from Germany and 11th graders from the U.S. were included. In Germany, $n = 775$ participating students were from schools in North Rhine-Westphalia; 12.6 % of the students were from *Hauptschule*, 16.4 % from *Gesamtschule*, 26.6 % from *Realschule*, and 44.4 % from *Gymnasium*. The U.S. sample included $n = 528$ students from schools in the Chicago area. Teachers of all U.S. classes were involved in a teacher training on NOS and NOSI.

8   Data Analysis

Data analysis in this thesis focuses on the validation of the devised competence model (cf. Chapter 4). To sufficiently represent this model, a large amount of NOSSI test items needed to be developed. Asking students to solve all of these items (i.e. more than 100 per student) would have been overstraining and not feasible. Therefore, students were given only subsets of items. Commonly, classical test theory (CTT) is used to analyze test data. Yet, CTT methods for NOSSI data analysis would only allow for inferences for par- ticular subsets of items but not on all items together. In contrast, probabilistic test theory (PTT), Rasch measurement in particular, provides measures for persons' abilities and items' difficulties as if all students had responded to all items. Consequently, NOSSI test items were analyzed based on a Rasch analysis. Those instruments that had not been administered in subsets (i.e., KFT, LGVT, SNOS and SIS_{ad}) were analyzed based on CTT. The basic assumptions of CTT are summarized in Section 8.1. Section 8.2 illustrates the main idea of PTT in general, and Rasch analysis in particular. Finally, Section 0 describes the statistical procedures of inferential statistics that were used to further analyze item difficulty- and personal ability parameters obtained by Rasch measurement.

8.1 Classical Test Theory

In the studies presented here, several instruments were employed to determine particular traits of the surveyed students. For instance, the KFT was used to determine students' nonverbal intelligence, while the SIS_{ad} was employed to survey their physics related interest and self-beliefs (cf. Section 7.2). All instruments used consist of items, and all data gathered are the raw responses to these items. From these raw responses, a student's value of a trait was inferred (e.g., his or her degree of intelligence or degree of interest in physics). A test theory was used to describe and explain the relation between a specific trait of a person and his or her performance on a test that was designed to determine this trait (cf. Rost, 2004). In other words, a test theory is necessary to infer trait values from raw responses on an instrument. Typically, classical test theory is used to make such inferences (cf. Bortz & Döring, 2006; Bühner, 2006; Haertel, 2006; Lienert & Raatz, 1998).

CTT assumes that a true trait value cannot be measured exactly. Rather, a measured score regarding a specific trait is composed by a person's true value regarding this trait and a measurement error. The measurement error is assumed to average at zero when the trait is repeatedly measured. As a consequence, taking the mean of repeatedly measured scores reveals the true trait value according to CTT (cf. Bortz & Döring, 2006).

Therefore, instruments usually contain more than one item to measure a particular trait to keep the measurement error small. For example, the SIS_{ad} scales contain four to seven items to determine particular aspects of interest and self-beliefs. Typically, Cronbach's reliability coefficient α is utilized to explore how well such items of one scale measure the same trait. Cronbach's α therefore is one, if not *the* important CTT criterion for instrument quality (cf. Lienert & Raatz, 1998). The α coefficient indicates to what extent an instrument, or particular scales of an instrument, is internally consistent (i.e. to what extent the single item responses intercorrelate to each other). Higher α-values indicate a higher internal consistency of an instrument or particular scales. Typically, $\alpha \geq .70$ or $\alpha \geq .80$ is viewed as indicating a high reliability (cf. Field, 2009).

8.2 Probabilistic Test Theory

Like Classical test theory, probabilistic test theory (PTT) – Rasch modeling in particular – provides an approach to make inferences on traits based on students' responses to items (e.g., Bond & Fox, 2007; Rost, 2004; Wilson, 2005; Yen & Fitzpatrick, 2006). Rasch modeling assumes so-called specific objectivity which means that the revealed Rasch measure of a person's ability is independent from the set of items the person responded to (Rost, 2004). Accordingly, persons' abilities concerning the construct itself can be estimated as well as the probabilities of a correct response to the rest of the items for which no observed responses are available. Conversely, item difficulties can be estimated as if all persons responded to all items. Consequently, even in multi-matrix designs, Rasch measurement takes into account all items and all persons for gaining measured of item difficulty and person ability, which CTT does not.

Therefore, data gathered with the NOSSI items (Section 7.1) were analyzed by means of Rasch modeling. Here, the simplest model of the family of Rasch models was applied. In this model, a person's performance on a test item is expressed as a probability that is based on item difficulty and person's ability[17]. Such a probabilistic relationship between latent trait and performance on an item is in contrast to CTT's assumptions, which represent a direct relationship between trait and performance.

Rasch analysis that was employed in the presented studies contained two main steps. First, NOSSI data were fit to Rasch models of different dimensionality to identify a best fitting model that could be used as a starting point for further analyses. Then, items were investigated regarding their fit to this chosen model. All Rasch analyses were performed using the program CONQUEST 2.0 (Wu, Adams, & Haldane, 2007). For time efficiency with respect to the estimation of higher dimensional models, *Monte Carlo* estimation with *1000 nodes* was used. Constraint was set to items so that mean item difficulty was fixed at zero for each dimension. Termination criterion for parameter change and deviance change was set to 10^{-4}, the maximum number of iterations was set to 3000.

Comparing Rasch models

A specific feature of the Rasch model is unidimensionality. Once it is ensured that the data fit the Rasch model, unidimensionality can be assumed; that means that all items then represent one latent trait (cf. Bond & Fox, 2007). Accordingly, the use of different Rasch models allows for conclusions to be drawn about the structure of the latent trait(s) to be measured. In particular, it is possible to investigate if two (or more) sets of items correspond to two (or more) traits (e.g., if the NOSSI items correspond to a unified NOSI/NOS competence or if one item subset corresponds to NOSI competence and the

---

[17] More complex Rasch models also include chance level and discrimination which may vary between items.

other one to NOS competence). Therefore, Rasch models of different dimensionalities were investigated here to examine the number of traits represented by the different *complexity* levels (and *cognitive processes* and *core aspects*, respectively). In doing so, Rasch models of different dimensionality are compared to each other with an *n*-dimensional model being a simultaneous estimation of *n* unidimensional models.

To be able to decide which of the compared models fits the data better, the *deviance* statistic is utilized. The *deviance* statistic expresses "how well the item response model has fit the data" (Wu, Adams, & Wilson, 2007, p. 23). The smaller the *deviance* of an estimated model, the better the fit between the data and estimated model is. Since *deviance* and, thus, the difference between *deviances* is χ^2-distributed, a χ^2-test can be used to check if the difference in *deviance* of two models is statistically significant with the difference in the number of estimated parameters corresponding to the degrees of freedom of χ^2-distribution (cf. Wu et al., 2007; Rost, 2004). This test of significant difference can only be used if the models to be compared are so-called 'nested' models. This means that one model has to emerge from the other model by restricting parameters; however, parameters may not be constrained to zero (Rost, 2004). For example, when comparing a unidimensional model with a two-dimensional model – i.e., when comparing unified NOSI/NOS competence vs. separate NOSI and NOS competences – the two-dimensional model is nested in the unidimensional one; this is because in the two-dimensional model, all items of the unidimensional one are assigned a particular dimension, but none is excluded. All comparisons that are made here involve only nested models. Moreover, validity of the non-restricted model has to be shown. Another requirement for using this test refers to the probability of observed patterns. Only if all possible response patterns show an expected frequency of 1, or more simply put, only if every possible response pattern is represented in the data matrix, can the deviance statistic be assumed to be χ^2-distributed. It can be assumed that this prerequisite is met if a large sample per item is available (cf. Rost, 2004).

Another approach to compare models is based on information theory (Rost, 2004). In line with Occam's razor, simplicity of the estimated model is taken into account. In this case, simplicity is represented by the number of estimated parameters n_p. Rost (2004) lists the following three coefficients:

- Akaike's Information Criterion (AIC): $AIC = deviance + n_p$
- Bayes' Information Criterion (BIC): $BIC = deviance + n_p \cdot (logN)$
- Consistent Akaike's Information Criterion: $cAIC = deviance + n_p \cdot (logN) + n_p$

These coefficients can be used for model comparisons if the conditions for a comparison of *deviance* are violated – e.g., if the assumption of *deviance* being χ^2-distributed is violated. However, Rost (2004) points out that there is no rule delineating what difference is sufficient for choosing one model over another.

For the data analysis here, *deviance* of the models was used to compare models of different dimensionality. Using a comparison with χ^2-distribution, differences in *deviance* are investigated for statistical significance. In this case, a type I error may not exceed the critical value of $\alpha = .05$. Additionally, information coefficients were used. However, only BIC and cAIC were used because AIC is not appropriate for large samples.

Model fit

After comparing models of different dimensionality and after deciding for one particular model, items are investigated with respect to their fit to the model. So-called "fit statistics" are used to "detect the discrepancies between the Rasch model prescriptions and the data we have collected in practice" (Bond & Fox, 2007, p. 235). Since person and item measures are used for further analyses, only those items fitting the Rasch model should be included; otherwise, values of these measures could be skewed and lead to wrong conclusions of the further analyses. CONQUEST provides *outfit* and *infit mean square statistics* (abbreviated *outfit* and *infit*) to identify misfitting items. In contrast to *outfit*, *infit* is a statistic weighted by variance. Both, *outfit* and *infit*, indicate the relation between observed and expected variance in response patterns. Moreover, the standardized forms of these two *mean square statistics* are provided: So-called *outfit-T* and *infit-T* are then normalized on the sample size. In addition to *mean square statistics* CONQUEST also provides a classical indicator for item quality: so-called *discrimination*. *Discrimination* equals the point-biserial correlation calculated by the correlation between the persons' score on a particular item and their sum score, and thus indicates an item's power to distinguish between persons. Items with low *discrimination* are not useful because they do not contribute any information (cf. Lienert & Raatz, 1998).

The question of cutoff limits of fit indices also has to be thoroughly discussed. *Mean square statistics* and *T-values* depend on sample size. Bond and Fox (2007) pointed out that deciding based on mean square statistics, "we are likely to declare that all items fit well when the sample size is large enough" (p. 241ff) while deciding based on *T-values*, "we are likely to reject most items when the sample is large enough" (p. 243). Literature on Rasch analysis suggests different cutoff limits. Wright (1996, cited by Smith, Schumacker, & Bush, 1998) suggested smaller upper limits for *mean square statistics* the larger the sample size is. Wilson (2005) referred to Adams and Khoo (1996) who "have indicated that .75 (=3/4) is a reasonable lower bound and 1.33 (=4/3) is a reasonable upper bound" (Wilson, 2005, p. 129) for *infit* statistics. Bond and Fox (2007) indicated [0.75; 1.3] as general acceptability interval for *mean square statistics* and [-2.0; 2.0] as general acceptability interval for *T-values*. Rost (2004) used the common 95%-limits of standard normal distributed *T-values* and specifies [-1.96; 1.96] as acceptability interval. Investigating other projects that employed Rasch analysis reveals varying cutoff limits, too. For example, from the technical reports of the Programme of International Student Assessment (PISA), a non-uniform picture can be found. Analyzing data from the PISA 2000 study, dodgy items were detected by $discrimination < .25$ and/or $infit \notin [.80; 1.20]$ (Adams, 2002). The PISA 2003 study categorized all items showing $discrimination < .25$ were categorized as dodgy (OECD, 2005), while PISA 2006 applies the weaker criterion of $discrimination < .20$ (OECD, 2009). When analyzing data of the Third International Mathematics and Science Study (TIMSS), cutoff limits were set with $infit \in [.88; 1.12]$ and $discrimination < .20$ (Mullis & Martin, 1998). This overview shows that finding a guideline in terms of cutoff values is quite hard. Bond and Fox (2007, p. 241) pointed out that the "interpretation of fit statistics […] requires experience related to the particular measurement context" and suggested considering that "omitting the overfitting items […] could rob the test of its best items".

For the data analysis here, *discrimination, weighted mean square statistics (infit)*, and *infit-T* were analyzed for each item to detect misfitting items. The following acceptability cut-off values were used:

- $discrimination \geq .20$
- $infit \in [.80; 1.20]$
- $infit\text{-}T \leq 2.0$.

These chosen cutoff criteria are moderate, but not careless. For subsequent item parameter analysis especially, the number of items should not be extensively decreased so as to not limit possible effects. The use of only one cut-off – instead of an interval – for *infit-T* was chosen because low *infit-T* values indicate items that fit the Rasch model too well. Excluding those items would mean excluding the best items.

Since persons and items are treated equivalently in Rasch analysis, the same fit statistics could be used to identify 'misfitting' persons. On the one hand, excluding persons from data analysis could lead to an improvement of model fit, but on the other hand, this would mean reducing the sample size and limiting generalizability. Accordingly, the misfitting of person parameters was neglected when analyzing the data here.

8.3 Procedures of Inferential Statistics

To investigate the research questions and hypotheses, the item and person parameters obtained by Rasch analysis were further analyzed using inference statistics. Hypotheses concerning the internal structure of the competence model's three dimensions were explored by employing analyses of variance. Hypotheses concerning discriminant validity were investigated based on the correlations of competence regarding NOSI and NOS on the one hand, with competence-related variables on the other. In contrast to the cognitive ability test KFT and the reading comprehension/speed test LGVT, the two Likert-type questionnaires on beliefs about the nature of science (SNOS) and on physics-related interest and self-beliefs (SIS$_{ad}$) are not standardized. These questionnaires were analyzed using a confirmatory factor analysis before their relationship with the NOSSI test was explored. For comparing two subsamples with respect to their NOSI/NOS competence (curricular validity), a t-test was used. In the following, the basic principles of these procedures of inferential statistics are described.

Analysis of variance

Analysis of variance (ANOVA) is usually used to compare more than two groups regarding a particular trait. A comparison of each pair of groups using several t-tests would serve to reach this goal, too, however, it would result in an inflation of the Type I error (cf., e.g., Bortz, 2005; Field, 2009). ANOVA requires normally distributed data within each group and homogeneity of variance between groups. It enables the identification of a significant effect on a particular trait, but also makes possible the investigation of differences between those groups that were hypothesized previously (so-called 'planned contrasts', Field, 2009). The so-called 'explained variance' η^2 indicates the size of the effect. Cohen (1988, 1992) specifies (1) small effects by $\eta^2 \in [0.01; 0.06[$, (2) medium effects by $\eta^2 \in [0.06; 0.14[$, and (3) large effects by $\eta^2 \geq 0.14$.

One-way ANOVA – this is the case of one factor's effect on one variable – was used to identify the effects of each, *complexity level, cognitive process*, and *aspect*, on NOSSI item difficulty. To ensure normal distribution of data and homogeneity of variance, a Kolmogorov-Smirnov test and Levene's test were employed. To investigate if *complexity levels* were separated, contrasts were explored. The same method was used to examine the differentiation between *cognitive processes*. Explained variance η^2 was used to evaluate the effect size.

Confirmatory factor analysis

Confirmatory factor analysis (CFA) was used to analyze data gathered with SIS$_{ad}$ and SNOS questionnaires to identify which variables might be reasonably used for correlational analyses with person parameters regarding NOSSI[18]. In particular, CFA was expected to help to determine whether using each scale separately or using combined scales would be more appropriate.

---

18 Investigating dimensionality using Rasch analysis is similar to conducting a CFA. Unfortunately, Rasch analysis could not be employed to explore the structure of SIS$_{ad}$ and SNOS questionnaires; there was no robust Rasch model found which the data fitted. In these cases, CFA was employed.

Well-founded and theoretically derived models are central prerequisites for implementing CFA. CFA enables investigating the quality of fit of one or more (competing) models to a data set at hand. The basic principle of CFA is a comparison of the observed data set and the hypothesized model regarding their variance-covariance matrices or correlation matrices, respectively (cf. Bühner, 2006). Based on the difference between these two matrices, a significant discrepancy between the data and the hypothesized model can be determined by using a so-called χ^2 goodness of fit test. Additionally, the use of other fit parameters is suggested for model evaluation (Beauducel & Wittmann, 2005). Root Mean Square Error of Approximation (RMSEA) takes into account discrepancy, sample size, and numbers of observed and estimated parameters. Comparative Fit Index (CFI) relates an estimated model to a so-called independence model that assumes all parameters to be set at zero. RMSEA is to be preferred to CFI if missing values have to be estimated (cf. Bühner, 2006). Beauducel and Wittmann (2005) also suggested including Standardized Root Mean Residual (SRMR). However, SRMR can only be determined for data sets without any missing value and, therefore, cannot be used for the data set at hand. As in the case of Rasch modeling, Akaike's Information Criterion (AIC) can be determined for models estimated using CFA. In this case, AIC can be derived by $AIC = \chi^2 + 2 \cdot n_{ep}$ with n_{ep} being the number of estimated parameters.

In terms of evaluating these fit criteria, the χ^2 goodness of fit test should reveal a non-significant χ^2 value indicating an exact model fit. Particularly in the case of a significant χ^2 value and a large sample size, Bühner (2006) suggested evaluating additional fit indices (as CFI and RMSEA). He recommended cut-off criteria of $RMSEA \leq 0.06$ for $n > 250$, and $CFI \approx 0.95$. As in the case of Rasch analysis, small AIC values are preferable.

As already pointed out above, a prerequisite for performing a CFA is to have reasonable theories or empirically plausible assumptions to test for. Additionally, data have to meet some criteria. At best, data to be analyzed should be at an interval level of measurement and normally distributed. Depending on the estimation algorithm, normal distribution of data is more or less necessary (for an overview on different estimation methods, see Bühner, 2006). Maximum likelihood method of estimation (ML) is suggested to be preferred to all other methods because of its robustness against violation of normal distribution. Bühner (2006) pointed out that such violation often results in excessively high χ^2 values.

CFAs were performed using the program AMOS (Arbuckle, 2009). For each questionnaire, SIS_{ad} and SNOS, competing models were assumed and compared regarding their fit to data. Method of estimation was ML, because (1) of its robustness against non-normal data, and (2) because ML allows for the estimation of means in cases of missing values: This ensures keeping a sample as large as possible. Moreover, competing models were compared by evaluating χ^2 value, CFI, RMSEA, and AIC.

Correlations

Employing correlations is a common method for investigating criterion-related validity (e.g., Best & Kahn, 2006; Bortz, 2005; Furr & Bacharach, 2008). Based on the data gathered in both studies (Sections 7.2 and 7.3), Pearson's product-moment correlation r was employed in order to investigate the relation between competence on NOSI and NOS on the one hand, and cognitive ability, reading speed and comprehension, beliefs about the nature of science, and interest and self-belief scales, on the other. To evaluate the effect size, the following intervals suggested by Cohen (1988) are used: $r \in [.10; .30[$ indicates a small effect, whereas $r \in [.30; .50[$ can be considered a medium, and $r \geq .50$ a large effect.

To identify the correlation between item characteristics, like *complexity* level or the *cognitive process* an item refers to, Pearson's correlation coefficient cannot be applied, since the assignment of *complexity* levels and *cognitive processes* is only ordinal. Therefore, Spearman's non-parametric correlation coefficient r_s is used for such analyses. For large sample sizes, power estimates do not seriously differ between Pearson's product-moment correlation r and Spearman's correlation coefficient r_s (Cohen, 1988). Consequently, to identify effect size for r_s, the thresholds used for r can be adopted.

Student's t-test

In order to compare two subsamples concerning one trait, Student's t-test for independent samples was used (Bortz, 2005; Field, 2009). To obtain the t-statistic, the difference in the means between two subsamples is related to the weighted average variance of the total sample. The independent t-test provides the information if two samples show significantly different means concerning a particular variable. Accordingly, using the t-test is appropriate for identifying differences in NOSI and NOS competence between German and U.S. students. The difference is viewed as significant if the α-level is less than .05. To estimate the effect size, the observed t-value can be converted into an r-value. As in the case of correlations $r \in [.10; .30[$ indicates a small effect, whereas $r \in [.30; .50[$ refers to a medium, and $r \geq .50$ refers to a large effect.

Results

The main research question of this work concerns the validity of the model of competence regarding Nature of Scientific Inquiry and Nature of Scientific Knowledge. The corresponding hypotheses refer to aspects of internal and external validity (Chapters 0 and 6). Data were gathered after the theoretical competence model was operationalized into items (NOSSI test, Section 7.1). A subsequent Rasch analysis of the NOSSI-test data provided measures of item difficulty and person ability (Section 8.2). In principle, item difficulties were used to investigate aspects of internal validity, whereas aspects of external validity were investigated based on person abilities. The two conducted studies included both, investigation of internal and external validity aspects. In Chapter 9 the results from Study 1 are detailed, and in Chapter 10 the findings from Study 2 are. Finally, these results are discussed in Chapter 11.

9   Results of Study 1

The first study was conducted to examine (1) the internal structure of the model of competence (hypotheses H1, H2, and H3) and (2) the differentiation between competence on NOS and NOSI on the one hand, and cognitive ability, reading ability, beliefs about the nature of science, and interest and self-beliefs concerning physics on the other hand (hypothesis H4). In terms of data analysis, this means that the NOSSI test was analyzed from two different perspectives: first, item parameters (items' difficulties), and second, person parameters (students' abilities). In Section 0, the internal structure of the model of competence is investigated. This is followed by a discussion of the relations between NOSI and NOS competence and competence-related external constructs in Section 9.2.

9.1 Internal Structure of the Model of Competence

Hypotheses H1, H2, and H3 share a similar structure: First, data have to be analyzed regarding the number of traits (sub-hypotheses H1a, H2a, and H3a), after which factors influencing item difficulty need to be examined (sub-hypotheses H1b/c, H2b/c, and H3b). To investigate these hypotheses, only NOSSI test data are used. NOSSI data were coded dichotomously (correct/wrong). In cases where students gave no answer or answered ambiguously, items were coded "wrong". These dichotomous data were fitted to Rasch models of different dimensionalities in order to investigate the number of traits the items represented. Afterwards, influences on item difficulty were investigated using analyses of variance (ANOVA).

Comparing models of different dimensionality

The underlying idea of investigating Rasch models of different dimensionality (cf. Section 8.2) shall be discussed exemplarily using hypothesis H1a: '*Complexity* represents a unidimensional variable.' First, the type of Rasch dimensionality used in this investigation is *between item dimensionality* because items are clearly assigned <u>one</u> level of *complexity* (the same holds true for *cognitive processes*, and concept, NOSI or NOS). If every level of *complexity* corresponded to a separate latent trait a five-dimensional Rasch model would be fitting: In this case, each of the Rasch dimensions would correspond to one level of *complexity* and the five subgroups of items would correspond to five separate latent traits. However, the data fitting a unidimensional Rasch model indicate that there is only one latent trait represented by all items. Thus, a unidimensional Rasch model with a better fit than a five-dimensional Rasch model would corroborate the assumed internal consistency of *complexity levels*. The structure of *cognitive processes* and *concepts* has been similarly examined. Data were fitted to the following five models:

- Unidimensional model (1-dim): This model fitting indicates that items represent one latent trait ("competence in NOS and NOSI").

- Two-dimensional model (2-dim): This model fitting indicates that a subsample of the items represents one latent trait ("competence in NOS") and the other items represent another latent trait ("competence in NOSI").

- Six-dimensional model (6-dim): This model fitting indicates that six subsamples of the items represent six latent traits. Each trait represents competence regarding one core aspect of NOS and NOSI.

- Four-dimensional model (4-dim): This model fitting indicates that four latent traits are represented by four subsamples of the items, with each trait standing for one cognitive process separate from the others.

- Five-dimensional model (5-dim): This model fitting indicates that five latent traits are represented by five subsamples of the items, with each trait standing for one level of complexity separate from the others.

Table 4 shows which of these models were compared. Additionally indicated are hypotheses that correspond to these comparisons and their related expectations. Criteria used for such comparisons were *deviance* statistics, Bayes' Information Criterion (BIC), and consistent Akaike's Information Criterion (cAIC) as described in Section 8.2.

Table 4. Comparisons of the Different Rasch Models Conducted

Hypothesis	Compared Models	Expectation
H1a (*complexity*)	1-dim vs. 5-dim	1-dim fits best
H2a (*cognitive processes*)	1-dim vs. 4-dim	1-dim fits best
H3a (*NOSI and NOS*)	1-dim vs. 2-dim vs. 6-dim	2-dim fits best

As illustrated in Section 4.3, *complexity* represents a unidimensional trait; in particular, *complexity* levels are expected to represent different qualities of one single trait (hypothesis H1a). Thus, a unidimensional Rasch model is expected to show a better fit than a five-dimensional model. Since the unidimensional model emerges from the five-dimensional one by restricting parameters (i.e., the models are nested; cf. Section 8.2), deviance statistics can be used as a criterion for model comparisons. Table 5 shows that the *deviance* of the unidimensional model is higher than the *deviance* of the five-dimensional model. Given $df = 14$ degrees of freedom, this difference of $\Delta_{dev,1/5} = 8.00$ is not significant (Section 8.2). Both information coefficients, BIC and cAIC, indicate a preference for the unidimensional model. Thus, hypothesis H1a *cannot* be rejected.

Analogously, *cognitive processes* are expected to represent four grades of one single trait. Comparing the unidimensional and the four-dimensional model (Table 5) reveals that the four-dimensional model fits significantly better than the unidimensional model ($\Delta_{dev} = 20.01$; $df = 9$; $p < .05$) – in contrast to the expectation. However, the four-dimensional model comes off worse with respect to both information coefficients, BIC and cAIC. Because there is no unambiguous conclusion to be inferred from these criteria, hypothesis H2a *cannot* be rejected. Further investigations seem necessary to elucidate this aspect more thoroughly.

In order to explore the relation between the two concepts, NOS and NOSI, data were fitted into a one-, a two- and a six-dimensional Rasch model. Based on the above-illustrated theory (Section 3.1), NOSI and NOS are expected to be separated constructs (two-dimensional model), but aspects are not (six-dimensional model). Estimation of the six-dimensional model showed convergence problems. This means that the difference in deviance between two subsequent iterations did not become smaller than the convergence criterion (here 10^{-4}, cf. Section 8.2). Consequently, convergence problems of the six-dimensional model here indicate that this model is not robust and, thus, not appropriate for further analyses. A comparison of *deviances* reveals that the unidimensional and the two-dimensional models do not significantly differ ($\Delta_{dev} = 0.68$; $df = 2$; ns). Both information coefficients show better values for the unidimensional model. In summary, hypothesis H3a *cannot* be rejected because criteria of comparison do not clearly show a preference for one model. The latent correlation between the two dimensions was high, $r_{latent} = .89$.

Table 5. Comparison of Rasch Models Regarding Deviance Statistics, and Information Coefficients BIC and cAIC (Study 1).

model	deviance	# parameters	BIC	cAIC
1-dim	17453.01	107	18200.38	18307.38
5-dim	17445.01	121	18290.16	18411.16
4-dim	17433.00	116	18243.23	18359.23
2-dim	17452.33	109	18213.67	18322.67
6-dim	not robust*			

*Convergence problems during estimation.

To sum up the above three model comparisons, two central aspects shall be highlighted. First, in none of these comparisons did data allow for clear rejection or corroboration of hypotheses. Thus, these aspects should be investigated once again (e.g., in Study 2). Second, since the hypothesis of the two-dimensional fitting could not be rejected and since the two-dimensional model was in line with the theory of nature of science, the two-dimensional model was chosen as the basis for further analyses.

Investigating influences on items difficulty

The model of competence on NOSI and NOS (Chapter 4) contains assumptions regarding factors generating items' difficulty. *Complexity* levels (H1) as well as *cognitive processes* (H2) are hypothesized to show an effect on item difficulty and to be ordered hierarchically. In contrast, aspects of NOSI and NOS are assumed to have no influence on item difficulty (H3). In order to examine these assumptions, effects on item parameters were analyzed using a one-way ANOVA approach (Section 0).

As the two-dimensional Rasch model was chosen as a starting point for further analyses, items were examined concerning their fit to this model before analyzing effects on item parameters. Based on *discrimination*, *infit*, and *infit-T* (cf. Section 8.2), nine items were excluded. The remaining 97 items were included in analyses of variance. Prerequisites for conducting an ANOVA, normal distribution of data and homogeneity of variance were checked. None of the groups that were investigated using ANOVA significantly differed from normal distribution. Additionally, Levene's test revealed homogeneity of variances for all variables to be investigated: *complexity* levels, *cognitive processes*, and *aspects* of NOS and NOSI (a detailed overview of both tests' findings is given in Appendix C.1).

Hypotheses with respect to *complexity*'s influence on item difficulty take on two aspects. First, a significant effect of complexity on item difficulty is expected (H1b); second, mean difficulty per *complexity* level is assumed to increase strictly monotonically (H1c). *Complexity* levels were significantly correlated to item difficulty, $r_s = 0.32$, $p < .01$. One-way ANOVA revealed that the *complexity* level has a large effect on item difficulty, $F(4,92) = 3.84$, $p < .01$, $\eta^2 = .14$. Figure 8 depicts means and confidence intervals for each level of complexity. Based on an analysis of contrasts, levels which can be significantly distinguished from each other were indicated: Level I differed from Level III, IV, and V; Level II differed from Level III, and V. Means M of each level are listed in Table 6. They do not meet the assumption of strict monotony; rather, Level III and IV showed a disorder. In summary, hypothesis H1b is corroborated, and hypothesis H1c is partly corroborated.

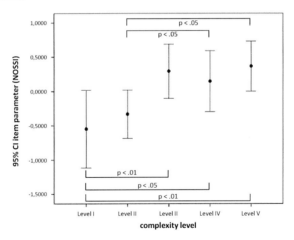

Figure 8. Mean NOSSI item difficulties per *complexity* level (Study 1). Significant group differences are indicated (two-tailed).

Table 6. Mean Item Difficulty $M(SD)$ per Complexity Level (Study 1).

Complexity Level	I	II	III	IV	V
$M(SD)$	$-0.55\ (1.14)$	$-0.33\ (0.73)$	$0.29\ (0.89)$	$0.15\ (0.92)$	$0.37\ (0.75)$

Similarly, *cognitive processes* are expected to have a significant effect on item difficulty (H2a) and to be ordered hierarchically; that is, mean difficulty per *cognitive process* increases strictly monotonically (H2b). *Cognitive processes* and item difficulty were significantly related, $r_S = 0.38$, $p < .01$. Moreover, they showed a large effect on item difficulty, $F(3,93) = 6.12$, $p < .01$, $\eta^2 = .17$. Means and confidence intervals for each *cognitive process* are displayed in Figure 9. An analysis of planned contrasts revealed that reproducing differs from organizing and integrating, and that selecting differs from integrating. Examining the mean item difficulties (Table 7) for each cognitive process revealed that they are ordered according to the hypothesis. In summary, these results corroborate the hypotheses H2b and H2c.

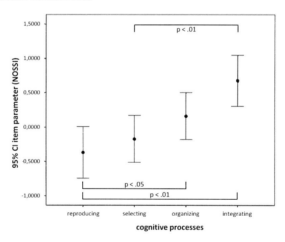

Figure 9. Mean NOSSI item difficulties per *cognitive process* (Study 1). Significant group differences are indicated (two-tailed).

Table 7. Mean Item Difficulty $M(SD)$ per Cognitive Process (Study 1).

Cognitive process	reproducing	selecting	organizing	integrating
$M(SD)$	$-0.37\ (1.03)$	$-0.17\ (0.87)$	$0.16\ (0.73)$	$0.68\ (0.77)$

In principle, an interaction effect of *complexity* and *cognitive processes* on item difficulty could be possible. To investigate such interaction, a two-way ANOVA would have to be calculated. Groups included in this analysis were defined by each combination of *complexity* level and *cognitive process*. However, in this case, Levene's test indicated unequal variances ($F(15,81) = 1.85$, $p < .05$). Taking into account that the group sizes differ between 4 and 10, assumptions for calculating an ANOVA were violated (cf. Bortz, 2005; Field, 2009). Thus, an interaction effect could not be investigated based on this data set.

Aspects of NOSI and NOS, respectively, were assumed to not influence item difficulty (H3b). Conducting a one-way ANOVA revealed that *aspects* hold no significant effect on item parameters, $F(5,91) = 0.45$, $p = .82$. These results corroborate hypothesis H3b.

Summing up these results, effects of *complexity* level and *cognitive processes* on item difficulty could be determined, whereas *aspects* showed no influence. This is in line with the hypotheses. The assumed hierarchical order of *cognitive processes* could be confirmed. However, the order of *complexity* levels was shown only by trend because Level III and Level IV were disordered. In both cases, *cognitive processes* and *complexity* levels, gradations were not fully separated from each other.

9.2 Relations to Control Variables

To investigate aspects of external validity, competence regarding NOSI and NOS was compared to other closely related cognitive variables – (1) nonverbal cognitive ability, (2) reading speed and reading comprehension, (3) interest and self-beliefs, and (4) beliefs about the nature of science (hypothesis H4). A large correlation between NOSI/NOS competence and these external variables would indicate a close interrelation. As a consequence, such a large correlation would be a sign of NOSI/NOS competence not being distinguishable from another variable. Accordingly, hypotheses H4a/b/c/d would have to be rejected if the correlations show a large effect, i.e. if $r \geq .50$ (cf. Section 0). Hypothesis H4e refers to a comparison of these correlations. It is assumed that the beliefs concerning the nature of science are more closely related to competence regarding NOSI and NOS than the other cognitive variables. Consequently, correlations between NOSI/NOS competence and the SNOS scales are expected to be larger than those between NOSI/NOS competence and the other cognitive variables (i.e., nonverbal cognitive ability, reading speed/comprehension, interest and self-belief scales).

As explained in the preceding section, the two-dimensional Rasch model was chosen as a starting point for further analyses. Additionally, misfitting items were removed. The final estimation included 97 items that did not show misfit anymore. Before further analyses were made, NOSI and NOS person parameters were checked for normal distribution. Kolmogorov-Smirnov tests revealed that NOS person parameters showed normal distribution ($Z(1080) = 1.19$, $p = .12$), while NOSI parameters did not ($Z(1080) = 1.66$, $p < .01$). Hence, to identify the central measure, mean would have to be used for NOS parameters, while the median would be appropriate for NOSI parameters. To keep comparability, median of both NOSI and NOS parameters are reported here: median for NOSI was found to be $P_{50} = 0.81$ ($min = -3.84$; $max = 3.62$) and for NOS to be $P_{50} = 0.85$ ($min = -3.41$; $max = 3.52$). Because even small effects tend to become significant with large sample sizes, the non normal distribution of the NOSI parameters might be due to the sample size and, thus, could be neglected. Comparing median and means for both NOSI and NOS parameters shows that median and mean are very close to each other (cf. Appendix C.2). Yet strictly speaking, medians are appropriate for non-normally distributed data; therefore, the median is further used.

	Persons		Items
4	NOS	NOSI	

```
                                                X|
                            X|
 3                         XX|                    X|
                            X|                    X|
                           XX|                  XXX|
                           XX|                   XX|
                         XXXX|                   XX|
                          XXX|                 XXXX|
                        XXXXX|                 XXXX|
                    XXXXXXXXX|             XXXXXXXX|58  90
 2                     XXXXXX|               XXXXXXX|
                      XXXXXXX|             XXXXXXXXX|19
                   XXXXXXXXXX|           XXXXXXXXXX|31
                 XXXXXXXXXXXX|         XXXXXXXXXXXX|66  93
                XXXXXXXXXXXXX|        XXXXXXXXXXXX|
              XXXXXXXXXXXXXXX|      XXXXXXXXXXXXX|50
             XXXXXXXXXXXXXXX|  XXXXXXXXXXXXXXXX|20
            XXXXXXXXXXXXXXXX|    XXXXXXXXXXXXXX|10
           XXXXXXXXXXXXXXXXXXXXXXXXXXXXXXXXXXXXXX|41  46  74  97
 1  XXXXXXXXXXXXXXXXXXXX|   XXXXXXXXXXXXXXXXX|1  12  60
    XXXXXXXXXXXXXXXXXX|    XXXXXXXXXXXXXXXXXX|95
      XXXXXXXXXXXXXX|       XXXXXXXXXXXXXXX|9  48  92
    XXXXXXXXXXXXXXXXXX|    XXXXXXXXXXXXXXXX|37  42  57  67  69  78  85  88  91  94
   XXXXXXXXXXXXXXXXXX|      XXXXXXXXXXXXXX|4  18  35  81
      XXXXXXXXXXXX|       XXXXXXXXXXXXX|25  54  83
        XXXXXXXXXXXX|       XXXXXXXXXXXX|43  65  87
       XXXXXXXXXXXX|      XXXXXXXXXXXX|2  72  89
 0     XXXXXXXXXXXXX|         XXXXXXXXXX|8  23  28  30  51  53  68  76
         XXXXXXXXXXX|         XXXXXXX|5  29  34  56  73  82
            XXXXXXX|          XXXXXX|49  55
            XXXXXX|           XXXXX|3  24  27  79
            XXXXXX|            XXXX|61
            XXXXXX|          XXXXXX|14  22  44  52  80
               XX|             XX|6  45  59  62  77  86
              XXX|            XXX|33  75
               XX|              X|7  16  26  39  64  70
-1            XX|              XX|13  21  47  84
               X|               X|36
                |               X|38  71
                |                |11  17  32  96
              X|                |15
              X|
              X|
-2              |
                |
                |
                |
                |
                |40  63
```

Figure 10. Wright Map of the two-dimensional Rasch model based on NOSSI items (Study 1). Each 'X' represents 3.3 students.

In the case of the used Rasch estimation, *expected a posteriori/plausible values reliabilities* (EAP/PV) for each dimension were $r_{EAP/PV}(NOSI) = .55$ and $r_{EAP/PV}(NOS) = .55$. Comparable to Cronbach's α, EAP/PV reliability indicates how consistently person parameters could be measured. A Wright map (person-item-map, Figure 10) displays the distributions of item parameters and person parameters. Note that in the case of a two-dimensional Rasch model, person parameters are provided for both dimensions. Figure 10 reveals that item and person distributions are shifted: Mean item parameter is constrained to zero, while median for NOSI was $P_{50} = 0.81$ and for NOS was $P_{50} = 0.85$. This means that items were too easy for this sample. The observed shift is one reason for relatively low reliability values. Person-item match is inappropriate; adequate items are missing, especially at the high end of the scale. In each used NOSSI booklet, the minimum number of items per dimension was 5 items. This relatively low number could be another reason for low reliability values; the more items per dimension a student's response is available for, the more precisely his or her ability can be measured.

Nonverbal cognitive ability

The nonverbal scale N2 of the cognitive ability test KFT (Heller & Perleth, 2000; cf. Section 7.2) was employed using two versions. N=538 students worked on test version A, n=529 students on test version B. Raw data of KFT (nonverbal scale N2) were coded dichotomously (correct/wrong). In cases where students gave no answer or answered ambiguously, items were coded "wrong". In accordance with the KFT manual, a student's total was transformed into standardized T-values making results from the different forms (A and B) comparable. If students showed a total below the minimum specified T-value, they were given this minimum T-value. Reliability was satisfying ($\alpha = 0.83$ for version A, and $\alpha = 0.85$ for version B). T-values (all students) were not normally distributed (cf. Appendix C.3). The median[19] was found to be $P_{50} = 37$, and observed minimum and maximum t-values were $min = 25$ and $max = 63$.

To investigate the relation between competence regarding NOSI and NOS on the one hand, and nonverbal cognitive ability on the other hand, Pearson's correlation coefficient r between person parameters (for both dimensions, NOSI and NOS) and the T-values of KFT was determined. Nonverbal cognitive ability was found to correlate with competence regarding NOS, $r = .27$, $p < .01$, and with competence regarding NOSI, $r = .25$, $p < .01$. Both relations show a small effect and, thus, corroborate hypothesis H4a.

---

[19]As in the case of NOSI and NOS person parameters, median was chosen as the measure of central tendency here. Means and median of non-verbal cognitive abilities are close to each other (cf. Appendix C.3).

Reading abilities

To identify reading comprehension, students' responses to a cloze test were analyzed (cf. Section 7.2). In accordance with the *Lesegeschwindigkeits- und –verständnistest[20]* LGVT test manual (Schneider, Schlagmüller, & Ennemoser, 2007), filling options were coded 1 for a correct answer, 0 for no answer, -1 for a wrong or ambiguous answer. Based on the manual, totals were transformed into standardized T-values. Reading speed was determined based on the number of words read. Again, totals were transformed into standardized T-values. Sixty-seven cases occurred, in which no reading speed could be determined because the students did not indicate how much they had read.

Since the LGVT test is a standardized instrument and because Cronbach's alpha cannot be determined based on available data, reliability values were accessed from the manual. There, retest reliability was indicated to be $\alpha = .87$ regarding reading comprehension, and $\alpha = .84$ regarding reading speed (Schneider, Schlagmüller, & Ennemoser, 2007). For both reading speed and comprehension, observed T-values were not normally distributed. The median[21] of reading comprehension (N=1063) was $P_{50} = 51$; minimum and maximum observed values were $min = 17$ and $max = 78$. The median of reading speed (N=996) was $P_{50} = 52$; minimum and maximum observed values were $min = 29$ and $max = 76$. A detailed distribution is shown in Appendix C.4. Reading speed and reading comprehension showed a correlation of $r = .59, p < .01$.

Table 8 displays the results of comparing the T-values concerning reading comprehension and reading speed, as well as competence regarding NOSI and NOS. Reading speed tends to be less related to both competence regarding NOSI and regarding NOS than reading comprehension does. However, all correlations do not exceed a medium effect. This result corroborates hypothesis H4b.

Table 8. Pearson's Correlation Coefficient r for Reading Comprehension/Speed and Person Parameters.

Person Parameter	NOS	NOSI
$r_{\text{reading comprehension}}$ $(n = 1063)$	.29, $p < .01$	.27, $p < .01$
$r_{\text{reading speed}}$ $(n = 996)$	.15, $p < .01$	.10, $p < .01$

Note. n indicates the number of students included in the analysis.

---

[20] Reading speed and reading comprehension test
[21] As in the case of NOSI and NOS person parameters, median was chosen as the measure of central tendency here. Means and median of each reading speed and reading comprehension are close to each other, too (cf. Appendix C.4).

Interest and self-belief

To control for aspects of interest and self-belief regarding physics, the Self-Belief and Interest Questionnaire for students was employed using a version adapted for 10th graders (SIS_{ad}, see also Section 7.2). SIS_{ad} consisted of six scales: situational interest in physics lessons (FI), out-of-school interest in physics (AI), personal interest in physics (SI), experienced competence in physics (EK), self-efficacy in physics (SW), and self-concept regarding physics (SK). The first three scales can be reasonably subsumed under interest, while the latter three scales can be grouped under self-belief (cf. Kauertz, 2009). However, the simplest model would assume that only one factor underlies all six scales. To identify the structure underlying SIS questionnaire, a confirmatory factor analysis (CFA) was conducted. Three models were compared: (1) a one-factorial model assuming all items load on one factor; (2) a two-factorial model assuming that FI-, AI-, and SI-items load on one (e.g., interest), and EK, SW, and SK-items load on a second factor (e.g., self-belief); and (3) a six-factorial model assuming that items of each scale load on a factor separate from all items of the other scales. As described in Section 0, AMOS (Arbuckle, 2009) was used to perform the ML estimation. Models were compared regarding χ^2, CFI, RMSEA, and AIC coefficients. Table 9 shows the results of the three models to be compared.

Table 9. Comparison of Fit Indices Revealed from a One-, Two-, and Six-factorial CFA of SIS$_{ad}$ Questionnaire.

	One-factorial	Two-factorial	Six-factorial
χ^2 (df; p)	6115.9 (527;.000)	4162.7 (526;.000)	1940.0 (512; .000)
RMSEA	0.100	0.080	0.051
CFI	0.741	0.831	0.934
AIC	6319.9	4368.7	2174.0

Table 9 reveals that none of the models showed an exact fit (χ^2 values are significant). In reference to Bühner (2006), this indicates small misspecifications of the models. Regarding RMSEA, CFI, and AIC, the six-factorial model is to be preferred to the other two models. Moreover, only concerning the six-factorial model RMSEA and CFI show acceptable values. Based on these results, the six scales of SIS$_{ad}$ are separately used for subsequent analyses. Assuming the six-factorial model is the most appropriate also reveals the most detailed results for further analyses.

As a consequence, the means for each scale and person were determined. Calculating means was seen to be appropriate in cases where a person responded to at least $n_S/2 + 1$ items (for an even amount of items per scale n_S) or $(n_S + 1)/2$ items (for odd n_S). In all other cases, determining a mean scale value is seen to be too imprecise. Table 10 shows statistical values regarding the six scale means as well as Cronbach's α for each scale. Appendix C.5 provides a detailed overview of the distributions of the means per scale within the sample. Kauertz (2009) reports reliabilities between 0.71 (experienced competence) and 0.86 (situational interest, self concept) concerning the original questionnaire, which was employed in a sample of 4th- and 6th graders. Comparing these reported values to the observed reliabilities, the questionnaire's adaptation obviously worked out well – the adapted version's performance was satisfying.

Table 10. Statistics of SIS$_{ad}$ Scales: Median[22] of Scale Means P_{50}, Sample Size n and Cronbach's α.

Scale	P_{50}	n	$\alpha\,(n_{items}, n_{sample})$
Situational interest (FI)	2.14	1067	0.87 (7, 1033)
Out-of-school interest (AI)	1.67	1062	0.83 (6, 1051)
Personal interest (SI)	1.60	1065	0.89 (5, 1060)
Experienced competence (EK)	2.50	1069	0.81 (4, 1067)
Self-concept (SK)	2.43	1064	0.88 (5, 1059)
Self-efficacy (SW)	2.40	1068	0.90 (7, 1035)

Correlations between the scale means of SIS$_{ad}$ and person parameters were investigated to clarify the relation between competence regarding NOSI and NOS and interest and self-belief. The results presented in Table 11 show that situational interest (FI) and out-of-school interest in physics (AI) were not significantly correlated with competence regarding NOSI and NOS. All other scales –personal interest in physics (SI), experienced competence (EK), self-concept (SK), and self-efficacy (SW) concerning physics – showed a significant relationship to NOSI and NOS competence. However, the effects of each relationship remained small. These results are in line with hypothesis H4c.

Table 11. Pearson's Correlation Coefficient r for Scale Means of SIS$_{ad}$ and Person Parameters.

Scale	FI	AI	SI	EK	SK	SW
n	1067	1062	1065	1069	1064	1068
$r(NOSI)$	.01	.003	.06*	.10**	.17**	.18**
$r(NOS)$	.02	.03	.09**	.08**	.17**	.19**

*$p < .05$ (two-tailed); **$p < .01$ (two-tailed); n indicates the number of students included in analysis
n indicates the number of students included in the correlation for each scale

---

[22]Due to scale means being not normally distributed, median was chosen as the measure of central tendency here. Means and median of each SIS$_{ad}$ scale yet are close to each other (cf. Appendix C.5).

Beliefs about the nature of science and science-related epistemological beliefs

To investigate the relation between beliefs about the nature of science and competence regarding NOSI/NOS, the Seven Scales on the Nature of Science Questionnaire (SNOS) was employed (Section 7.2). The SNOS questionnaire consists of seven scales: Source (Q), Certainty (S), Development (E), Justification (R), Simplicity (K), Purpose (Z), and Creativity (V). A Confirmatory Factor Analysis (CFA) was used in order to determine if a combined score for all seven scales could be reasonably used for further analyses of the relation between beliefs about the nature of science and competence regarding NOSI/NOS, or if seven separate scale scores should be used. Concerning those two possible approaches, a one-factorial model was compared to a seven-factorial model. Similar to the procedure concerning SIS_{ad}, AMOS (Arbuckle, 2009) was used for ML estimation and to determine χ^2, CFI, RMSEA, and AIC coefficients according to which the models were compared. Table 12 shows the results of the three models to be compared.

Table 12. Comparison of Fit Indices Revealed from a One-, and Seven-factorial CFA of SNOS Questionnaire.

	One-factorial	Seven-factorial
χ^2 (df; p)	5334.5 (1274; .000)	3199.2 (1253; .000)
RMSEA	0.055	0.038
CFI	0.502	0.762
AIC	5646.5	3553.2

As in the case of SIS_{ad}, none of the models showed an exact fit (χ^2 values are significant). RMSEA values of both models meet defined cut-off criteria, while CFI values do not. However, CFI tends to be better for the seven-factorial model than for the one-factorial model. AIC also indicated preference for the seven-factorial model over the one-factorial model. Additionally, as the basis for further analyses, a seven-factorial model will reveal more detailed results than having a one-factorial model. In summary, the seven-factorial model was preferred to the one-factorial model and was used as a starting point for further analyses.

Comparable to the procedure concerning SIS_{ad}, the means for each scale and person were determined in cases where a person responded to at least $n_S/2 + 1$ items (for an even amount of items per scale n_S) or $(n_S + 1)/2$ items (for odd n_S). Statistical values regarding the seven scale means as well as Cronbach's α for each scale are shown in Table 13. Appendix C.6 provides a detailed overview of the distributions of the means per scale within the sample.

The total reliability of the questionnaire was found $\alpha = .81$. Urhahne, Kremer, and Mayer (2008) reported a total reliability of $\alpha = .84$ and scale reliabilities between $\alpha = .52$ (Simplicity) and $\alpha = .71$ (Development). The observed statistical values and reliabilities of each scale are listed in Table 13. Similar to the reported reliabilities, 'Simplicity' shows the lowest, and 'Development' the highest reliability. Based on these results, the SNOS questionnaire performed similarly satisfying. The reliability of the 'Simplicity' scale ($\alpha = .47$) is quite low. However, this scale was not changed because (1) the validity of the scale should be kept and (2) removing one item would have led to only a marginal increase of reliability (of 0.004).

Table 13. Statistics of SNOS Scales: Median[23] of Scale Means P_{50}, Sample Size n and Cronbach's α.

Scale	P_{50}	n	$\alpha\ (n_{items}; n_{sample})$
Source (Q)	4.17	1066	.72 (6; 1044)
Certainty (S)	3.75	1065	.63 (8; 1017)
Development (E)	4.00	1065	.73 (8; 1019)
Justification (R)	4.09	1064	.61 (11; 1017)
Simplicity (K)	3.00	1066	.47 (5; 1041)
Purpose (Z)	3.86	1064	.51 (7; 1024)
Creativity (V)	3.29	1066	.70 (7; 1022)

Table 14. Pearson's Correlation Coefficient r for Scale Means of SNOS and Person Parameters.

Scale	Q	S	E	R	K	Z	V
n	1066	1065	1065	1064	1066	1064	1066
$r(NOSI)$	.22**	.23**	.28**	.23**	.17**	.11**	.07*
$r(NOS)$	.24**	.25**	.34**	.25**	.23**	.09**	.10**

$*p < .05$ (two-tailed); $**p < .01$ (two-tailed); n indicates the number of students included in analysis
n indicates the number of students included in the correlation for each scale

Means were determined for each scale of the SNOS questionnaire in order to investigate the relationship between beliefs about the nature of science and competence regarding NOSI and NOS. These means were correlated with the NOSI and NOS person parameters. Table 14 lists the observed correlation coefficients. Each scale was found to cor-

---

[23]Due to scale means being not normally distributed, median was chosen as the measure of central tendency here. Means and median of each SNOS scale yet are close to each other (cf. Appendix C.6).

relate with NOSI and NOS competence, however, the correlations range only from very small (NOS/SNOS-Z, and NOSI/SNOS-V) to medium (NOS/SNOS-E) effects. Hypothesis H4d is corroborated by these results.

Comparison of Correlations

Beliefs about the nature of science were expected to correlate with NOSI and NOS competence more strongly than the other external variables (hypothesis H4e). In order to investigate this assumption, 99.9% confidence intervals were determined for each correlation; this interval equals a 95% confidence interval including a Bonferroni correction for alpha error cumulation. Correlation coefficients, which had been revealed from the relation between NOSSI test and SNOS questionnaire (Table 14), were then compared to those revealed from the remaining correlations (nonverbal cognitive ability, reading comprehension/speed, scales of interest and self-belief). In doing so, competence regarding NOSI and regarding NOS were investigated separately. This means, for example, that the correlation between NOS competence and beliefs on sources of knowledge (SNOS-Q) was compared to the correlation between NOS competence and self-efficacy (SW); likewise, the correlation between NOSI competence and beliefs SNOS-Q was compared to the correlation between NOSI competence and SW. If the confidence intervals overlapped, correlations were viewed as being not significantly distinct from each other. If the confidence intervals were separated, the effect size of the relationship was compared. Confidence intervals of all significant correlations as well as an overview of the comparison of correlations are detailed in Appendix C.7. Because situational interest in physics lessons (FI) and out-of-school interest in physics (AI) were not significantly correlated with NOSI and NOS competence, these correlations were not included in the comparison.

A comparison of the significant correlations revealed that most correlations do not significantly differ (Table C 9 and Table C 10). Concerning NOSI competence, one comparison showed the expected direction (i.e., the correlation of NOSI competence with an SNOS scale was higher than those with another variable); concerning NOS competence, two comparisons showed the expected direction. These comparisons corroborate hypothesis H4e. However, the correlation of NOSI competence with the SNOS scale 'creativity (V)' as well as those of NOS competence with the SNOS scale 'purpose' (Z) were found to be smaller than the correlations with reading comprehension. This is probably because 'purpose (Z)' and 'creativity (V)' were found to be only weakly correlated with NOSI competence and NOS competence, respectively (cf. Table 14). As a consequence, hypothesis H4e *cannot* be rejected.

In summary, each pair of correlations between NOSI and NOS competence and a third variable (e.g., NOSI/EK and NOS/EK) tended to be of similar size. All correlations were found to not exceed a medium size. Accordingly, a distinction between NOSI and

NOS competence each and the investigated competence-related variables could be fully corroborated (hypotheses H4a/b/c/d). However, a comparison of the correlations (hypothesis H4e) revealed an unclear picture as to how much stronger beliefs about the nature of science and science-related epistemological beliefs correlated to NOSI and NOS competence are than the other variables.

10   Results of Study 2

Study 2 was conducted to serve two aims: (1) corroboration of findings concerning the competence model's internal structure, and (2) a comparison between German and U.S. students. Accordingly, the following results refer to data gathered in a sample mixed of German and U.S. students. Similar to the approach taken in Study 1, Rasch models of different dimensionalities and influences on item difficulty were investigated to reach the first aim. Descriptive results of this second study followed by an investigation of students' abilities concerning statistical differences between the two subsamples will be presented here.

10.1 Internal Structure of the Model of Competence

To corroborate the findings of Study 1 concerning the competence model's internal structure, Study 2 data were analyzed according to that of Study 1 by exploring hypotheses H1, H2 and H3. As in the case of Study 1, the number of traits was investigated by comparing Rasch models of different dimensionalities. Next, influencing factors on item difficulty were examined using analyses of variance (ANOVA). Since this approach is identical to the approach employed in Study 1, only the results will be presented in this section. More detailed information regarding the approach itself can be found in Section 0.

Comparing models of different dimensionality

Similar to Study 1, a unidimensional model (1-dim) was compared to a five-dimensional model (5-dim), a four-dimensional model (4-dim), a two-dimensional model (2-dim) and a six-dimensional model (6-dim) (Table 4). For these comparisons, *deviance* statistics as well as information coefficients BIC and cAIC (cf. Section 8.2) were used. Values of these criteria that were gained from Study 2 data are listed in Table 15.

Table 15. Comparison of Rasch Models Regarding Deviance Statistics, and Information Coefficients BIC and cAIC (Study 2).

Model	Deviance	# Parameters	BIC	cAIC
1-dim	22457.74	107	23225.19	23332.19
5-dim	22425.85	121	23293.71	23414.71
4-dim	not robust*			
2-dim	22446.01	109	23227.80	23336.80
6-dim	not robust*			

*Convergence problems during estimation.

Comparing the unidimensional model to the five-dimensional one (hypothesis H1a), *deviance* statistics revealed that the five-dimensional model has a better fit ($\Delta_{dev,1/5} = 31.89, df = 14, p < .05$). However, BIC and cAIC indicated a preference for the unidimensional model. As a consequence, hypothesis H1a *cannot* be rejected. Like in Study 1, *complexity* can thus be seen as one trait.

When fitting the data to the four-dimensional model, no convergent model could be found. This indicates that a four-dimensional model would not be robust and, thus, not appropriate for the data at hand. As a consequence, the unidimensional model could not be reasonably rejected in favor of the four-dimensional model, and thus, hypothesis H2a may *not* be rejected. Accordingly, *cognitive processes* may be viewed as a single trait, as was the case in Study 1.

Finally, the unidimensional model was compared to the two- and six-dimensional models (hypothesis H3a). Again, the six-dimensional model showed convergence problems and, thus, was not seen as robust. Comparing the *deviances*, the two-dimensional model showed a significantly better fit than the unidimensional model ($\Delta_{dev,1/2} = 11.73, df = 2, p < .05$). Similar to Study 1, BIC and cAIC of the unidimensional model were slightly smaller than for the two-dimensional one. Based on these results hypothesis H3a may *not* be rejected. As in the case of Study 1, the two-dimensional model was therefore chosen as a starting point for further data analyses.

Before further analyses were made, misfit of items was examined with respect to *discrimination, infit* and *infit-T* (cf. Section 8.2). Over nine cycles of excluding items and rerunning Rasch analyses on reduced item sets, twenty items had to be eliminated. After the last run, the remaining 86 items showed acceptable fit criteria. Based on the remaining items, the *expected a posteriori/plausible values reliabilities* (EAP/PV) for each dimension were $r_{EAP/PV}(NOSI) = .68$ and $r_{EAP/PV}(NOS) = .65$. Compared to Study 1, reliability increased and is now even more satisfying. Moreover, the Wright map (Figure 11) shows a better match between item and person parameters than in Study 1. Like in Study 1, mean item difficulty is constrained to zero. Person parameters on both dimensions were not normally distributed. The median regarding NOSI was found to be

$P_{50} = 0.25$ ($min = -3.47$; $max = 3.81$) and regarding NOS, to be $P_{50} = 0.38$ ($min = -3.01$; $max = 3.57$).

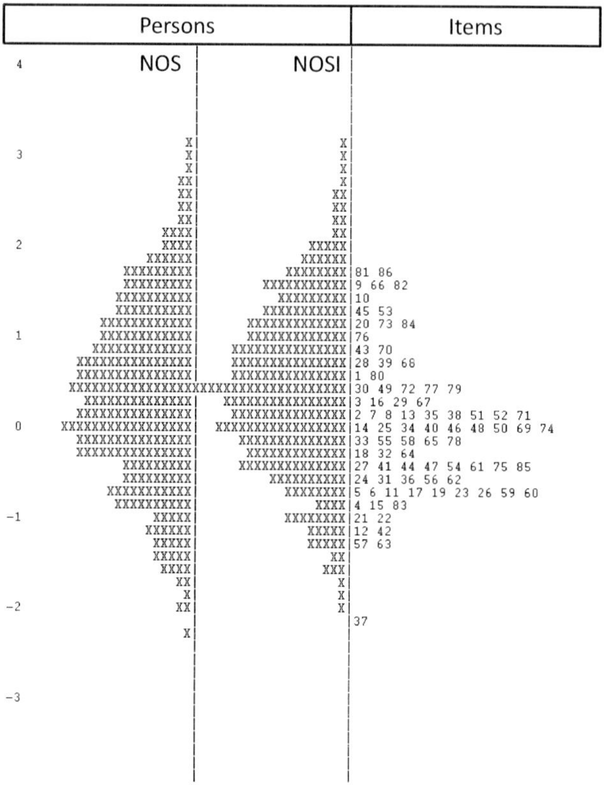

Figure 11. Wright Map of the two-dimensional Rasch model based on NOSSI items (Study 1). Each 'X' represents 4.5 students.

Investigating influences on items' difficulty

The remaining 86 items were included in analyses of variance. All groups that were investigated using ANOVA did not significantly differ from normal distribution. Additionally, Levene's test revealed homogeneity of variances for *cognitive processes* and *aspects* of NOS and NOSI (a detailed overview of both tests' findings is given in Appendix D.1). With respect to the *complexity* levels, Levene's test revealed unequal group variances, and thus indicated a violation of ANOVA assumptions. In this case however, ANOVA could be conducted because the group sizes (i.e. the number of items per *com-*

plexity level) were about the same (18, 16, 19, 16, and 17 items per group) and, thus, the F-statistic was fairly robust (cf. Field, 2009).

In Study 1, *complexity* had a significant effect on item difficulty (in line with hypothesis H1b). This result could not be corroborated by Study 2 data, $F(4,81) = 0.97$, $p = .43$. As a consequence, investigating contrasts was not reasonable in this case. However, *complexity* levels still correlated with item difficulty even if they did so to a lesser extent ($r_S = .23$, $p < .05$). Table 16 displays the mean item difficulties per *complexity* level. Again, the assumption of strict monotony (hypothesis H1c) was not met: Similar to Study 1, Level III and IV were disordered; moreover, Level I and II were flipped. When the items were grouped with respect to the three different elements of complexity – that is, facts, relations, and overarching concept – mean item difficulty increased. In summary, these results from Study 2 do not corroborate hypothesis H1b and only partly corroborate H1c.

Table 16. Mean Item Difficulty $M(SD)$ per Complexity Level (Study 2).

Complexity Level	I	II	III	IV	V
$M(SD)$	−0.16 (1.06)	−0.23 (0.69)	0.15 (0.82)	0.04 (0.59)	0.18 (0.62)
$M(SD)$		−0.20 (0.89)		0.10 (0.71)	0.18 (0.62)

The effect of *cognitive processes* on item difficulty (in line with hypothesis H2b), which was found in Study 1, could be corroborated in Study 2, $F(3,82) = 5.47$, $p < .01$, $\eta^2 = .17$. Even the relationship between *cognitive processes* and item difficulty increased ($r_S = .42$, $p < .01$). Means and confidence intervals for each *cognitive process* are displayed in Figure 12. The contrast analysis revealed the same differences that had been found in Study 1: Reproducing differed from organizing and integrating, and selecting differed from integrating. Study 1 also revealed that a strictly monotonic increase of difficulties per *cognitive process* (in line with hypothesis H2c) could be corroborated by the findings of Study 2 (Table 17). In summary, Study 2 revealed findings equivalent to Study 1 and substantiated hypotheses H2b and H2c.

Similar to Study 1, an interaction effect of *complexity* and *cognitive processes* on item difficulty could not be investigated based on this data set of Study 2. Levene's test showed that there was no homogeneity of variance between the different groups ($F(15,70) = 2.38$, $p < .01$). Moreover, the amount of items for each combination of *complexity* level and *cognitive process* were too different (between 1 and 11).

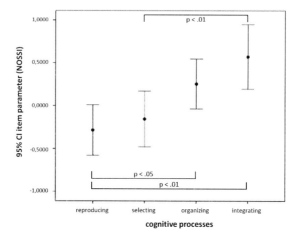

Figure 12. Mean NOSSI item difficulties per cognitive process (Study 2). Significant group differences are indicated (two-tailed).

Study 1 revealed that *aspects* of NOS and NOSI did not influence item difficulty (hypothesis H3b). Study 2 corroborated this finding ($F(5,80) = 0.31$, $p = .91$). Consequently, hypothesis H3b could be substantiated.

Table 17. Mean Item Difficulty $M(SD)$ per Cognitive Process (Study 2).

Cognitive process	reproducing	selecting	organizing	integrating
$M(SD)$	−0.29 (0.77)	−0.16 (0.80)	0.25 (0.57)	0.57 (0.65)

In summary, *cognitive processes* had an effect on item difficulty, while *aspects* did not – similar to Study 1. However, a *complexity* levels effect could not be observed. *Cognitive processes* were ordered hierarchically, but not fully separated from each other. *Complexity* levels were only ordered by trend, when Levels I and II ('facts') were consolidated, as well as Levels III and IV ('relations'), respectively.

10.2 Comparing German and U.S. Students

The comparison between the two subsamples of German and U.S. students should contribute to the curricular validity of the competence model (hypothesis H5). U.S. students were expected to outperform German students with respect to their NOSI and NOS competence (Chapter 6). Typically, to compare two samples to each other Student's t-test for independent samples is used. This approach, however, requires the data to be normally distributed, yet the data investigated here were not (Table 18). Therefore, the t-test's findings were double-checked by using the non-parametric Mann-Whitney-U test. The Student's t-test revealed that the subsamples differed concerning competence regarding NOSI ($t(1088.17) = 12.22$, $p < .01$, $r = .35$) and NOS ($t(1301) = 13.22$, $p < .01$, $r = .34$). A Mann-Whitney-U test corroborated the findings from the t-test, $U_{NOSI} = -11.45$, $p < .01$ and $U_{NOS} = -12.91, p < .01$. A comparison of the medians, however, revealed that German students performed better than U.S. students (Table 19). Consequently, hypothesis H5 has to be rejected.

Table 18. Results of the Kolmogorov-Smirnov-Test on the NOSI and NOS Person Parameters for Each Subsample.

	NOSI	NOS
German (n=775)	$Z(775) = 1.55, p < .05$	$Z(775) = 1.79, p < .01$
U.S. (n=528)	$Z(528) = 1.52, p < .05$	$Z(528) = 2.77, p < .01$

Table 19. Median of Person Parameters Concerning NOSI and NOS for Each Subsample.

Person Parameter (P_{50})	**NOSI**	**NOS**
German (n=775)	0.57	0.62
U.S. (n=528)	−0.28	−0.14

11   Discussion

The aim of this project was to establish empirical support for a competence model regarding Nature of Scientific Inquiry and Nature of Scientific Knowledge. A respective model was deduced from theory in Chapter 4. This model embraced three dimensions: *complexity*, *cognitive processes*, and *content*. The dimensions *complexity* and *cognitive processes* were assumed to determine levels of competence, while the *content* dimension addressed two components of the nature of science; namely, Nature of Scientific Inquiry and Nature of Scientific Knowledge.

Empirical support for this model was established by investigating construct validity and criterion validity. Construct validity was examined by exploring if the three model dimensions each met the assumed attributes. Criterion validity was investigated by discriminating NOSI and NOS competence against competence-related constructs. Curricular validity was explored by comparing a German and a U.S. sample that were expected to differ.

The assumed structure of the dimension complexity *is represented by the data (H1)*

The *complexity* dimension embraces five levels: one fact, two facts, one relation, two relations, and overarching concept. These levels were assumed (1) to represent *one joint* construct, (2) to influence test item difficulty, and (3) to be hierarchically ordered. These characteristics were investigated in both studies.

To explore whether the five levels define different manifestations of one construct or whether each of the levels corresponds to an individual construct, data were fitted to a unidimensional Rasch model and to a five-dimensional Rasch model, respectively. Model comparison was performed based on *deviance* statistics and two information criteria, BIC and cAIC. In both studies, no evidence supported choosing the five-dimensional models over the unidimensional models. Consequently, findings suggest that the five *complexity* levels represent *one* construct.

To identify the relationship between *complexity* levels and item difficulty, an analysis of variance was conducted, and Spearman's correlation coefficient was determined. Study 1 revealed *complexity* to have a medium effect on item difficulty; yet the levels could not fully be separated from each other: Level I did not significantly differ from Level II, Level II did not significantly differ from Level III and so on. The findings of Study 2 did not reveal any significant effect. However, both studies revealed that item difficulty is correlated with *complexity* levels, with a low to medium effect size. This finding indicates that on average, item difficulty increases with increasing *complexity* level. In Study 2, 20 of 106 items had to be excluded because they did not fit the Rasch model. This reduction of items might explain the missing effect in Study 2. However, the missing effect could also be due to the mixed sample of German and U.S. students. Study 2 showed that German students outperformed U.S. students. It is possible that the NOSSI test contained items that – compared to the other items – favored German students. Such differences might be the reason why no effect could be found in Study 2. Differences in item difficulty with respect to different samples could be investigated by analyzing differential item functioning (DIF), yet the research design did not allow for DIF analyses. In summary, the influence of *complexity* levels on item difficulty could be only partly corroborated.

To substantiate that *complexity* levels were hierarchically ordered, mean item difficulties were expected to monotonically increase with increasing complexity level. Study 1 revealed this overall trend as expected; however, Level III and IV were disordered. The findings of Study 2 confirmed this observed order; additionally, Level I and II were flipped. Aggregating facts (Level I and II) and relations (Level III and IV), the expected trend was shown: Mean item difficulty increased from facts to relations to the level of overarching concepts.

In summary, the results only partially substantiated hypothesis H1; *complexity* was not found to be an adequate construct to define levels of competence concerning nature of science. In contrast, Kauertz (2008) was able to show that nearly a third of variance in item difficulty of physics items was explained by *complexity*, yet the proposed six *complexity* levels did not exactly meet the expected order. Bernholt (2010) found *complexity* to explain more than half of the variance in difficulty of items on chemical combustion processes. Moreover, Bernholt was able to show that the levels meet the assumed hierarchical order.

The fact that Bernholt's findings are contrary to the findings here might lie in different operationalizations of *complexity*. Far more surprising is that the findings here are in contrast to the findings by Kauertz (2008), even if the operationalization of *complexity* developed from his work. The fact that the findings here are not in line with those by Bernholt (2010) and Kauertz (2008) might be because the items here unsatisfyingly operationalized the *complexity* levels. Further investigations therefore should include an expert rating with respect to the assignment of items to *complexity* levels. The result might also be due to the item content. Nature of science refers to the field of philosophy. In contrast, Bernholt's and Kauertz's work items focused on combustion processes and physics (i.e., on science content). *Complexity* was understood to determine different levels of complexity of information. It is possible that applying information about the nature of science does not become more difficult as information becomes more complex. Perhaps it is easier to understand and apply the concepts of nature of science (e.g., the core aspects) on a generalized level than to apply concrete relations. Moreover, students might always acquire knowledge, and thus competence, about the nature of science on a conceptual level. It also is possible that knowledge about the nature of science may not be structured in facts, relations and concepts and therefore would require another categorization. Consequently, other categorizations of knowledge structures should be explored with respect to the nature of science.

The assumed structure of the dimension cognitive processes *is represented by the data (H2)*

The *cognitive processes* dimension included four cognitive strategies: Reproducing, selecting, organizing, and integrating. These processes were assumed (1) to represent different gradations of the *same* construct, (2) to influence test item difficulty, and (3) to be hierarchically ordered. Both studies contributed to investigating these characteristics.

To explore whether the four processes derive from one construct or whether each of the processes corresponds to an individual construct, data were fitted to a unidimensional Rasch model and to a four-dimensional one. Model comparison was based on *deviance* statistics and two information criteria, BIC and cAIC. In both studies, no evidence was found to support choosing the four-dimensional models over the unidimensional ones. Consequently, the four *cognitive processes* can be justifiably viewed to represent *one* construct.

Analysis of variance as well as Spearman's correlation coefficient were employed to identify the relationship between item difficulty and *cognitive processes*. In both studies, *cognitive processes* were found to have a large effect on item difficulty. However, the four gradations could not be fully separated from each other: Reproducing did not significantly differ from selecting, selecting did not significantly differ from organizing, and organizing did not significantly differ from integrating. However, both studies revealed a medium effect size correlation between *cognitive process* and item difficultly; this means, on average, item difficulty increased with increasing *cognitive process*. In summary, the assumed effect of *cognitive processes* on item difficulty was substantiated by the findings of both studies. To support the assumed hierarchical order of the four *cognitive processes*, mean item difficulties were expected to monotonically increase with increasing *cognitive process*. The findings of both studies met this expectation. Consequently, *cognitive processes* justifiably make up a hierarchical order.

The findings above fully support hypothesis H2; *cognitive processes* might therefore be used to determine levels of competence regarding nature of science. A similar construct has already been employed by Kauertz (2008). So-called cognitive activities were included in Kauertz's model of physics competence to identify which cognitive requirements have to be mastered when solving an item. Cognitive activities were therefore based on the same underlying idea the *cognitive processes* are based on, unless the operationalization was different. According to Kauertz, cognitive activities were not appropriate for determining mastery levels of competence. In contrast, *cognitive processes* here were observed to be so. As in the case of hypothesis H1, such different results might stem from different operationalizations.

The *cognitive processes* used here are reminiscent of the cognitive processes proposed by Mayer (1984), who proposed that students successfully learn from a text through the sequence of selecting, organizing and integrating information. The presented studies showed that these processes – and the additional process of reproducing – can also be used to generate item difficulty, and thus to describe competence levels. With respect to teaching and learning about the nature of science by using historical contexts, these considerations suggest that students should first be challenged with reproduction or selection of information within a concrete context. By organizing such information, students should then be able to deduce more complex information. Finally, by extracting information from one context and integrating it into others, generalized concepts may be learned. With respect to item difficulty, such extraction is perhaps more important than the application of information of different *complexity*. However, it is unclear if the construct of *cognitive processes* is only appropriate in connection with historical contexts. In Section 3.2, abstract or inquiry-based approaches were named as other methods of teaching about the nature of science. What *cognitive processes* would be with respect to such contexts and if they would still influence item difficulty is still unclear and necessitates further research.

Competence regarding NOSI can be distinguished from competence regarding NOS (H3)

The *content* dimension embraces two components: Nature of Scientific Inquiry (NOSI) and Nature of Scientific Knowledge (NOS). Consequently, competence regarding NOSI was assumed to be a construct separate from competence regarding NOS. The core aspects of NOSI and NOS – in contrast to *complexity* levels and *cognitive processes* – were assumed to *not* influence the difficulty of an item. These characteristics of the *content* dimension were explored in both studies.

To investigate whether NOSI competence and NOS competence correspond to a joint construct or whether they correspond to two separate constructs, data were fitted to a unidimensional Rasch model and a two-dimensional one. In both studies, the models were compared based on *deviance* statistics, BIC and cAIC. The findings did not show any evidence for choosing the unidimensional models over the two-dimensional ones. Additionally, data were fitted to a six-dimensional model (assuming the core aspects of NOSI and NOS represent one construct each). Within both studies, the six-dimensional models were not robust, indicating that the data did not fit them, and thus, that the core aspects do not correspond to six separate constructs. Consequently, further analyses were based on the two-dimensional models. In both studies, competence regarding NOSI was highly correlated with competence regarding NOS. Conducting analyses of variance, no effect of core aspects on item difficulty was found in both studies. Obviously, the core aspects do not determine mastery levels.

Based on the findings, hypothesis H3 can neither be rejected, nor completely substantiated. The findings did not provide unambiguous evidence for NOSI competence and NOS competence as a joint construct. In theory, Nature of Scientific Inquiry and Nature of Scientific Knowledge are viewed as different concepts (cf. Lederman, 2007; Schwartz, et al., 2008; Section 3.1). Even if the findings here are ambiguous, they do not contradict this theoretical differentiation in terms of competences. A possible reason for the found ambiguity might be that the operationalization of the NOSI and NOS aspects through the items was perhaps not enough satisfying.

Assuming that NOSI competence and NOS competence are separate constructs, the latent correlation between these constructs was found to be quite high. One reason for that finding might lie in the ambiguity described above. If the items would have been unambiguously found to cover two dimensions, the correlation might have been lower. On the other hand, it also might be possible that NOSI and NOS are in fact separate constructs, but that they develop parallel to each other. It is possible that, by nature, the learning situations that focus on NOSI aspects are always related to NOS (ore vice versa). One reason might be that NOSI aspects determine NOS aspects (or vice versa). Another reason might be that students who understand a particular NOSI aspect be possibly able to also infer NOS aspects (or vice versa). As a consequence, if students have developed a competence regarding NOSI, they simultaneously might have also developed NOS competence (or vice versa). The presented studies did not aim to identify how NOSI and NOS competence develop; therefore, further investigations are needed to explore this development.

Competence regarding NOSI, and NOS, respectively, can be differentiated from other competence-related abilities and traits (H4)

To establish construct validity, competence regarding NOSI and NOS were discriminated against other competence-related constructs: (1) cognitive abilities, (2) reading speed and reading comprehension, (3) interest and self-belief regarding physics and physics education, as well as (4) beliefs about the nature of science and science-related epistemological beliefs. In Study 1, Pearson's product-moment correlation was employed to explore the relationship between competence regarding NOSI and NOS on the one hand, and the investigated competence-related constructs, on the other.

All correlations between competence regarding NOSI and NOS and competence-related constructs showed, at a maximum, a medium effect size as expected. Based on this finding, competence regarding NOSI and NOS can indeed be assumed to be constructs that are related to but still different from other competence-related abilities and traits.

Beliefs about the nature of science were expected to be more closely related to NOSI and NOS competence than the other constructs. However, most of the correlations were found to not differ. After correcting for alpha error cumulation, overall only three out of 98 possible comparisons revealed that the relation of NOSI or NOS competence with one of the seven belief scales was significantly higher than those relations with another construct. Two comparisons indicated a significant difference in the opposite direction: Competence regarding NOSI was less correlated with the SNOS-scale 'Creativity' than with reading comprehension; the same was observed for the correlation between competence regarding NOS and the SNOS-scale 'Purpose'. This is most likely due to the very small effect size ($r < .10$) of the correlation between 'Purpose' and 'Creativity' and competence regarding NOSI and NOS.

In summary, the findings largely corroborate hypothesis H4. Competence regarding NOSI and competence regarding NOS may therefore be viewed as independent constructs. In particular, the distinction against cognitive abilities indicates that NOSI and NOS competence are most likely not inborn; they might be assumed to be learnable, instead. Given the need to understanding about the nature of science with respect to *Allgemeinbildung* (cf. Chapter 1), NOSI and NOS therefore need to be covered in the scope of instruction. Students might otherwise not develop competence concerning NOSI and NOS. Likewise, NOSI and NOS competence were found to be different from reading abilities. This indicates that achieving competence regarding NOSI and NOS requires more than the ability to understand a text. NOSI and NOS competence's distinction from interest in physics and self-beliefs regarding physics means that there are students who have only low interest and self-beliefs with respect to physics but are competent with respect to NOSI and NOS. When dealing with NOSI and NOS topics in physics lessons, such students might feel successful and therefore might improve their self-belief and become more interested. Consequently, involving NOSI and NOS in physics lessons might foster students' interest in physics. Of course, this correlation would need to be investigated in further detail.

Interestingly, the relationship between NOSI and NOS competences and beliefs about the nature of science and science-related epistemological beliefs was not found to be as close as expected. A possible reason for that might be that the delineation of NOSI and NOS that was used in the NOSSI test did not exactly match the one used in the SNOS questionnaire. According to Mayer (2008), 'belief' is one of five types of knowledge. In contrast, competence refers to successfully applying knowledge in particular situations (Section 2.1); such knowledge would be – in Mayer's words – 'facts' and 'concepts', which refer to another type of knowledge. This difference could also have caused the relatively low correlations. To explore the relationship between beliefs about NOSI and NOS on the one hand and competence regarding NOSI and NOS on the other, further research is necessary. Such research could then also reveal if beliefs about NOSI and NOS need to be viewed as misconceptions and, therefore, if specific approaches of conceptual change (cf. Vosniadou, 1992) need to be employed.

Concerning NOSI and NOS competence, U.S. students show higher performance than German students (H5)

The topic of nature of science has a different significance in Germany and the U.S.: In the U.S. standards, this topic is mentioned explicitly, while the German standards only implicitly name it. As a consequence, U.S. students were expected to outperform German students. This expectation was explored in Study 2 investigating a German and a U.S. sample and by using Student's t-test and Mann-Whitney U-test. The two subsamples were found to significantly differ from each other. However, German students outperformed U.S. students. Thus, the results contradict the hypothesis H5, which therefore has to be rejected.

This surprising result is probably due to different test cultures in Germany and the United States. A pilot study, in which the U.S. students worked on the same booklets, and thus on the same amount of items per booklet as the German students, revealed that a lot of the U.S. students were less motivated than the German students (Section 7.3). Perhaps this was because the U.S. students were discouraged by the amount of text included in the items. Such discouragement could also have influenced the test results of the second study. It is also possible that tests in general are taken more seriously in Germany, or that German students are better acquainted with test items that require more reading than U.S. students are. This attitude towards tests in general and towards items that involve a lot of text in particular could have led to the unexpected performances.

This hypothesis's direction was based on the differences in the educational standards. However, the nature of science being explicitly considered in standards does not necessarily guarantee that it will be explicitly addressed in science lessons or that it will be addressed at all. Moreover, the study design did not include the gathering of the whole sample's personal background data (like intelligence, interest in physics, etc.). Those external variables could allow for the exclusion of an effect of nonequivalent samples (with respect to such constructs) – even if such variables obviously did not highly correlate with NOSI and NOS competence in a German sample (cf. findings from Study 1). In any event, this surprising result should be investigated in more detail. An intervention study in one country (e.g., students being explicitly taught aspects of the nature of science vs. those being regularly taught) should also be considered.

The validity of the competence model

The validity of the proposed competence model regarding NOSI and NOS was elucidated from different perspectives. To establish construct validity, hypotheses H1, H2 and H3 were investigated; hypotheses H4 and H5 provided insights into the model's criterion validity.

Investigating the structure of the model dimensions empirically was used to establish this aspect of validity. *Complexity* was not found to be an appropriate construct for determining competence levels in the field of nature of science. *Cognitive processes*, in contrast, seem to represent such construct; unfortunately, *cognitive processes* did not explain much variance in item difficulty. It is quite possible that, in fact, there is an interaction effect between *complexity* levels and *cognitive processes*. The combination of *complexity* levels and *cognitive processes* might provide another characteristic by which item difficulty is influenced and mastery levels of competence regarding NOSI and NOS can be identified. However, the investigated data sets did not allow for investigating such effect. In the presented studies, competence regarding NOSI could not be unambiguously differentiated from competence regarding NOS. Data did not allow for detailed analyses of the relationship between single core concepts, either. In summary, the findings did not provide an unambiguous substantiation of the dimensions' structures. Nevertheless, none of the findings indicated that the assumed structures need to be rejected. As a consequence, the model's construct validity could be partly established. Further research should investigate (1) if knowledge about the nature of science requires other operationalizations of *complexity*. (2) Studies should explore the interaction between *complexity* levels and *cognitive processes* with respect to their influence on item difficulty. (3) Moreover, other factors that generate item difficulty, such as the item's context, need to be identified. (4) The relation between NOSI and NOS competence needs to be investigated in further detail. (5) Finally, the core aspects and their interdependencies need to be investigated.

Concerning criterion validity, the presented results leave some open questions. Even if NOSI and NOS competence were empirically distinguished from other competence-related constructs, the expected proximity of NOSI and NOS competence to beliefs about the nature of science and science-related epistemological beliefs remained unclear and requires further research. Likewise, predictive validity remained unclear: German students' performed surprisingly well compared to U.S students. However, possible explanations of this result do not only include different abilities, but might also be due to other influences, like sample differences in motivation, average cognitive ability, reading abilities, attitudes towards science, experienced science lessons, out-of-school activities, or in acquaintance to the test type. All these variables should be controlled for in future studies that focus on comparing different samples. In summary, the findings concerning discriminant and predictive validity indicate that the model's criterion validity could be largely established.

Regardless of these open questions concerning construct and criterion validity, the devised model can be viewed as a viable starting point for research on competences regarding NOSI and NOS. This is of particular importance for German science education research. As illustrated in the theoretical background above, including the nature of science in German science education is necessary from the perspective of *Allgemeinbildung* (Chapter 1). From the perspective of current science lessons, Leisen (2009) demonstrated that – for the case of physics education – actual instruction is compatible with the teaching of nature of science aspects (cf. Section 3.2). However, clear guidelines of *what* and *how* to teach with respect to the nature of science are still missing. The present dissertation provides a starting point for this open question in a threefold way. First, a delineation of the concept of nature of science is provided. The description of NOSI and NOS by core aspects makes available a possible structure of this content, which is widely agreed on amongst science education researchers (cf. Section 3.1). By taking the concepts of NOSI and NOS as an extension of the German science education standards, a guideline of *what* to teach is provided. Second, the stories of physics history, upon which item development was based in this project, provide material for designing teaching and learning units on the nature of science. Opposing what is said in Leisen's appeal (2009), the literature on teaching approaches on the nature of science revealed the view that nature of science aspects have to be taught explicitly. Such explicit learning would also be in line with the idea of competences being learned in particular situations (cf. Klieme & Leutner, 2006). Finally, the developed instrument provides a large resource for test items, by which such teaching and learning units can be evaluated with respect to their efficacy. Despite what teaching approach is actually used (explicit, implicit, based on science history, based on inquiry, etc.), little research has been done on which of the approaches results in the best learning outcome. To provide more insight in this question, adequate instruments are needed. The items presented here can be used in such instruments – at least for the investigated age group.

Summary and Outlook

This dissertation's title suggests going 'Beyond Physics Content Knowledge'; and it does so in a two-fold manner: First, focusing on the nature of science means not to be restricted to physics content knowledge but to also reflect its characteristics; second, focusing on competences goes beyond the mere memorization of knowledge and instead refers to applying knowledge. Competence concerning the nature of science was argued to be a prerequisite for successful and responsible participation in science-related, decision-making processes. Without any doubt, appropriate teaching and learning strategies are necessary for the advancement of students' competences in this field. A starting point for such strategies is the unambiguous delineation of the learning goal, in this case, of competence regarding the nature of science. Such delineation is provided by competence models that detail the (sub-)competences of a particular domain and differentiate levels of mastery. Against the background of competence models, students' weaknesses can be precisely determined, and thus, specific and precise support can be provided to counteract them.

The goal of this project, therefore, was to theoretically devise and empirically validate a model of competence regarding Nature of Scientific Inquiry and Nature of Scientific Knowledge. The model that was derived from current research on competence and on the nature of science embraced three dimensions. The *complexity* dimension referred to different types of information. The dimension of *cognitive processes* addressed different cognitive strategies. The *content* dimension detailed the domain with two components of the nature of science, namely Nature of Scientific Inquiry and Nature of Scientific Knowledge. In order to investigate the empirical validity of the devised model, the model was operationalized into test items. Within two studies, the construct validity and criterion validity of the model was partly established.

Based on the findings of the studies conducted, several contributions to science education research are provided. From a practical point of view, a pool of test items surveying students' competences regarding NOSI and NOS is now available. Such items could be used to determine the efficacy of newly developed strategies for teaching aspects about the nature of science, for instance. From a theoretical standpoint, Nature of Scientific Inquiry and Nature of Scientific Knowledge were already assumed to be separate notions. With respect to competences, this assumption could only be observed as a trend; yet, it could not be empirically disproven either. For German science education, therefore, NOSI and NOS with their respective core aspects may serve as the theoretical basis for outlining curricula that should explicitly contain the nature of science.

Additionally, this dissertation project provides a theoretical, detailed definition of competence regarding NOSI and NOS. This defined competence could be empirically established as a new, independent cognitive construct. Furthermore, *cognitive processes* were shown to be a meaningful part of this competence construct. Even in an international sample, *cognitive processes* provided a reasonable basis for defining proficiency levels. And lastly, this project represents a first step in extending German science education standards – the model developed in this project is compatible with the model that has recently been developed to operationalize the science education standards. At present, these standards do not explicitly include nature of science aspects. Based on existing research literature, Nature of Scientific Inquiry and Nature of Scientific Knowledge were shown to be nature of science components that are important in school science contexts. This project showed what competences regarding these two components might reasonably be like. NOSI and NOS can, therefore, be used as a basis for extending German science education standards by explicitly including the domain of nature of science: For instance, they could be added as further basic concepts.

In addition to those theoretical and practical contributions, this project also raises several key questions for future research. First and foremost, the unclear picture that was observed concerning the *complexity* levels should be explored in more detail. Possible future directions for research might include an expert validation of the items' classification regarding the levels; improved definitions of the five levels; and a reduction to three levels only (fact, relation and concept). Closely related to this issue, a relatively small proportion of variance could be explained by *complexity* and *cognitive processes*. As a consequence, items need to be further analyzed to identify additional item characteristics binding variance. Two types of item characteristics should be analyzed. First, formal aspects of the items, like text length or readability should be investigated. If such formal aspects are found to bind large proportions of variance, items would be needed to be revised to rule out this influence. Second, characteristics that are related to students' abilities should be explored. Applying the concepts of NOSI and NOS in historical cases (as was the case here) could possibly relate to a different trait than applying them in abstract, philosophical contexts or with respect to inquiry processes would. Consequently, items at hand should be investigated with respect to whether the historical cases affect item difficulty or not; further items addressing other situations would then need to be developed and analyzed. Based on such analyses, the addressed situation could be identified as defining different sub-competences, and thus, should be included in the competence model as another dimension. Such revision could then also include the dismissal of the *complexity* dimension in favor of other, more decisive dimensions.

Second, the core aspects of NOSI and NOS could be investigated in more detail. Core aspects were shown to have no effect on item difficulty. However, it could be possible that some core aspects are closer to each other than others are. Such detailed analysis could provide helpful indications of how these aspects might be reasonably taught and learned. For example, aspects found to be closely related could be reasonably taught in concert with each other.

Third, the relationship between beliefs about the nature of science and competence regarding NOSI and NOS should be further investigated. In doing so, more parallel theoretical operationalizations of the investigated aspects might be used within both constructs – beliefs and competence. This way, the relationship between the presumably different cognitive constructs can be better explored, allowing researchers to rule out possible effects of the operationalizations.

Fourth, the surprisingly high performance of the German students, as compared to the U.S. students, has to be further examined. A further comparison study should include samples of other populations. Moreover, an intervention study appears to be reasonable; this way, possible unintended influences (e.g., due to different teachers or school curricula) on performances can be controlled. Moreover, background variables – like intelligence, beliefs about the nature of science – should be gathered in order to rule them out as unwanted influences as well.

Fifth, competence regarding NOSI and NOS should be assessed in the context of other science disciplines. The applied test items referred to only physics history. Contexts related to the other science disciplines could shed light on whether competence regarding NOSI and NOS is specific to the different disciplines or if it is specific to science in general.

Finally, strategies to implement nature of science aspects in school science lessons, in Germany at least, have to be developed and evaluated. Such material can be used to specifically enhance students' competence concerning the nature of science; a competence they will need for their future lives as responsible citizens. The historical contexts that were used for item development could provide a starting point for teaching- and learning material explicitly discussing nature of science aspects. Moreover, items provided here may be employed to evaluate newly developed teaching and learning units.

Appendix

A   Students' conceptions about the nature of science

The following table provides an overview of inadequate conceptions concerning nature of science, which could be identified from the reviews by Lederman (2007) and Höttecke (2001b, see also Höttecke, 2001a, 2004b). Höttecke (2001b) also includes some conceptions which are categorized to be adequate. Since this table is to illustrate the inadequate conceptions, those adequate ones are not listed here.

Table A 1. Inadequate Conceptions of the Nature of Science.

Inadequate conceptions summarized from Lederman's review (2007, pp. 836-838)

Students...
- think that scientific knowledge is absolutely and incontrovertibly true, i.e. they misjudge the tentative nature of scientific knowledge.
- think that the primary aim of science is to find out natural laws and the true facts.
- think that scientific theories mature into laws through constant testing and confirmation.
- think that scientific hypotheses can be verified.
- are not aware about how experiments, theories/models and the truth relate to each other.
- are not aware that scientific testing is important.
- do not know what makes up scientific explanations.
- do not sufficiently know about the role of scientific models.
- cannot distinguish between hypothesis, law and theory.
- underestimate subjective influences on scientific knowledge.
- underestimate the involvement of creativity within scientific knowledge.
- think that the purpose of science is to collect and classify facts.
- do not understand that scientific knowledge is parsimonious.
- misjudge that scientific knowledge is amoral, i.e. they do not understand that scientific knowledge cannot be used for moral judgments.
- do not comprehend that the different scientific disciplines, on the one hand, and the different types of scientific knowledge (hypotheses, laws, concepts...), on the other hand, contribute to a unified body of scientific knowledge.
- do not comprehend how the different branches of science relate and depend on each other.
- think that scientific knowledge is absolutely and incontrovertibly true, i.e. they misjudge the tentative nature of scientific knowledge.
- think that the primary aim of science is to find out natural laws and the true facts.
- think that scientific theories mature into laws through constant testing and confirmation.
- think that scientific hypotheses can be verified.
- are not aware about how experiments, theories/models and the truth relate to each other.
- are not aware that scientific testing is important.
- do not know what makes up scientific explanations.

- do not sufficiently know about the role of scientific models.
- cannot distinguish between hypothesis, law and theory.
- underestimate subjective influences on scientific knowledge.

Inadequate conceptions summarized from Höttecke's review (2001b, see also Höttecke, 2001a, 2004b)

1) The scientist as a person, the work of scientists, the conditions of scientific work
Students...
- view scientists as a stereotype: he is male; wears glasses, white coat, and beard (is unkempt); he works alone in a laboratory, isolated from the 'normal' world, being surrounded by research instruments, technology, formulas and books; he is highly intelligent; he can be dangerous (mad scientists), strange ("weird scientists", p. 9), helpful (solving problems), inventive concerning technical devices, or intellectual.
- think that scientists, on the one hand, are highly creative and, on the other hand, absolutely neutral when gathering data to find the truth about the nature.
- think that scientists' motivation to do research is noble and pure and due to individual thirst for knowledge (in contrast to reasons like earning money or respect within the scientific community).
- view public and common wealth (improving the world) as a purpose of science.
- view technology as a "product and area of application of science" (p. 11, translated by the author).
- are not aware that scientists are involved in other work than pure research (e.g., applying for funding, presenting at conferences, writing publications, supervising new researchers etc.).
- think that scientists are free of bias, i.e., they do not understand that they are pushed by success.

2) Scientific knowledge
Students...
- think that scientific knowledge involves "numbers, calculations, formulas and laws" (p. 13, translation by the author) whereas they underestimate the role of "private interests and considerations" (ibid.).
- think that natural laws identically reflect nature; accordingly, students think that "what is not known yet, has just not been investigated yet" (p. 13, translation by the author).
- think that scientific knowledge is unquestionably true.
- neglect the tentativeness and historicity of scientific knowledge, the influence of creativity, and the role of empirical evidence.
- cannot distinguish "between well-established and controversial knowledge" (p. 14, translation by the author).
- do not see the difference between evidence and explanation.
- think that a change in scientific knowledge leads to being closer to the truth.
- view scientific knowledge as being cumulatively developed.

3) Experiments as a part of scientific research
Students...
- have an insufficient understanding of how experiments and scientific knowledge relate to each other.

- underestimate the role of previous knowledge and expectations within the process of evaluating experimental data; this means they think of scientific investigations and measurements as being bias-free and objective.
- underestimate that experiments also involve "explorative and constructive aspects" (p. 17, translation by the author).
- misjudge the role of the scientific community concerning a) discussing, evaluating, or attaching importance/validity to particular experiments and b) following a particular paradigm (using particular types of experiments etc.).
- view experimenting as "trying out something and making discoveries" (p. 17, translation by the author).
- are not aware how hypothesis, experiment and data relate to each other.
- do not completely understand the role of measuring; they think e.g., that one measured value is correct, or that more precise measuring will reveal the true value.

4) Conditions of producing scientific knowledge
Students…
- think that the scientific community determines what is correct based on unambiguous data which show the truth.
- overestimate the evidence of hard facts and underestimate non-scientific influence (e.g., research programs, social factors etc.).
- view the sequence of question, hypothesis, data collection, and conclusion to be the scientific method; in particular, this sequence is seen as linear and unbranched.
- think that an experiment or an observation is the first step of scientific investigations.
- view repeated measuring as a means to decide on what is true.

B NOSSI Items

B.1 Item Distribution on Historical Contexts

Table B 1. Cases from Physics History used for Item Development.

Historical case	Physics topics	Period	# Items
Atomic Models	Atomic physics	20th century	6
Boyle's Law	Thermodynamics	17th century	9
The Energy Concept	Thermodynamics, Biophysics, Electromagnetism	19th century	9
The Universe	Astrophysics, Cosmology	20th century	11
Faraday's Experiment on Induction	Electromagnetism	19th century	10
The Composition of the Atomic Nucleus	Nuclear Physics	20th century	9
Newton and Colored Light	Optics	17th century	3
Nuclear Fission	Nuclear Physics	20th century	6
The Discovery of Electromagnetism	Electromagnetism	19th century	4
Ohm's Law	Electromagnetism	19th century	9
Superconductivity	Thermodynamics, Electromagnetism	20th century	13
The Solar System	Astronomy	4th century B.C. – 17th century A.D.	10
X-Rays	Atomic Physics	20th century	8

B.2 Item Distribution on Complexity Levels and Cognitive Processes

Table B 2. Item Matrix for developed Nature of Scientific Knowledge (NOS) Items.

NOS	Level I	Level II	Level III	Level IV	Level V	Σ_{items}
Integrating			3	2	5	10
Organizing		2	3	4	2	11
Selecting	5	4	3	3		15
Reproducing	5	3	4	3	3	18
Σ_{items}	10	9	13	12	10	54

Note: Only those items, which were included in data analysis, are listed.

Table B 3. Item Matrix for developed Nature of Scientific Inquiry (NOSI) Items.

NOSI	Level I	Level II	Level III	Level IV	Level V	Σ_{items}
Integrating			4	3	5	12
Organizing		5	2	2	2	11
Selecting	6	3	2	2		13
Reproducing	4	3	3	3	3	16
Σ_{items}	10	11	11	10	10	52

Note: Only those items, which were included in data analysis, are listed.

B.3 Developed Items

NOSSI items are listed according to their historical contexts. Sources for these contexts are indicated. Within each item, the correct option is checked. Additionally, *complexity level*, *cognitive process*, and core aspect of NOSI/NOS of each item are specified. Following abbreviations are used for aspects of NOSI/NOS:

NOS-1:   Scientific knowledge is influenced by subjectivity.

NOS-2:   Scientific knowledge is empirically based and inferential.

NOS-3:   Scientific knowledge is tentative.

NOSI-4:   Scientific investigations begin with a question.

NOSI-5:   Scientific investigations encompass multiple methods and approaches.

NOSI-6:   Scientific investigations allow multiple interpretations.

Finally, an overview on all items' location in the model is provided.

Atomic Models

Sources:

Geiger, & Marsden, 1909; Harré, 1984; Hermann, 1972; Rutherford, 1914; Schreier, 1984; Simonyi, 2004

Stem 1

Scientists conduct experiments and make observations, but they also develop theories and models. Theories, models, observations, and measured data have to fit each other. Theories and models are necessary to explain observations and measured data, and to generate hypotheses that lead to further investigations.

In the late 19th and early 20th century, scientists discovered the electron, X-rays and radioactivity. They then tried to devise a model of the atom in order to explain observations that occurred during their experiments. The model of the atom, which was suggested by Joseph John Thomson in 1903 (see picture 1), could be used to explain many observations. Therefore, this model was accepted by nearly all scientists. However, in 1909, the working group of the physicist Ernest Rutherford made observations that could not be explained with Thomson's model. It had taken Rutherford two years until he had developed a new model of the atom that could explain these observations (see picture 2), this model was accepted by the other scientists and Thomson's model had lost importance.

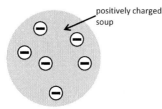

positively charged soup

picture 1: Thomson's model: atoms consist of a positively charged soup in which the electrons are embedded

picture 2: Rutherford's model: atoms consist of a positively charged nucleus which is surrounded electrons

Item AM1

Why were there two different models of the atom although both of them described the same thing?

☑   Scientific knowledge about the atom changed due to Rutherford's proposal.

☐   Thomson's model was wrong because it was based on the wrong data.

☐   Thomson did not read nature correctly and, thus, could not find the correct view.

☐   Rutherford's model gained acceptance because it truly described nature.

Complexity level	*Cognitive Process*	*Core aspect*
Level III	Reproducing	NOS-3

Item AM2

Why did Thomson develop a model of the atom and, thus, further scientific progress?

☑   In science, models are used to explain observations. Thomson proposed a model by which many observations could be described and explained.

☐   Scientists propose provisional models that lead to definite laws. Thus, Thomson proposed a model by which a law for X-rays could be inferred.

☐   Scientists investigate nature from which explanatory models emerge. Thomson's model was an hypothesis therefore leading to further investigations.

☐   In science, the knowledge is represented by models. Thomson's model represented the knowledge about atoms and, thus, could not be reversed.

Complexity level	*Cognitive Process*	*Core aspect*
Level V	Organizing	NOS-2

Item AM3

In which way did Thomson and/or Rutherford and his working group further scientific progress?

☐   Rutherford's working group worked experimentally and Rutherford devised his model based on these observations. Since Thomson developed his model only theoretically, it had to be refused by Rutherford's working group.

☐   Thomson's model could explain many observations. It was better because Rutherford's model was devised only to explain the observations his working group had made.

☑   Thomson and Rutherford developed models in order to explain observations. However, Rutherford's model was better because it also could be used to explain the observations his working group had made.

☐   Only Rutherford's working group collected data by observations. Thomson as well as Rutherford developed their models theoretically. Therefore, both of the models have to be seen as personal opinions.

Complexity level	*Cognitive Process*	*Core aspect*
Level IV	Reproducing	NOSI-5

Stem 2

Scientific knowledge is not definite and absolutely certain. It can change due to new observations and experimental outcomes. Additionally, new theories and interpretations can lead to modifications of knowledge.

In the late 19th and early 20th century, scientists discovered the electron, X-rays and radioactivity. They then tried to devise a model of the atom in order to explain observations that occurred during their experiments. The model of the atom, which was suggested by Joseph John Thomson in 1903 (see picture 1), could be used to explain many observations. Therefore, this model was accepted by nearly all scientists. However, in 1909, the working group of the physicist Ernest Rutherford made observations that could not be explained with Thomson's model. It had taken Rutherford two years until he had developed a new model of the atom that could explain these observations (see picture 2), this model was accepted by the other scientists and Thomson's model had lost importance. However, even Rutherford's model had limitations and had to be reworked successively. The model of the atom, that is still accepted today, is much more complex.

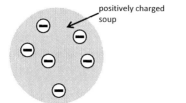

picture 1: Thomson's model:
atoms consist of a positively charged soup
in which the electrons are embedded

picture 2: Rutherford's model:
atoms consist of a positively charged
nucleus which is surrounded electrons

Item AM4

Why did Rutherford propose a model of the atom even if there already was one?

☑   His working group made observations and Rutherford wanted to explain them.

☐   Thomson had proposed a model and Rutherford wanted to prove it.

☐   Rutherford stated hypotheses on the atoms and tried to verify them.

☐   Rutherford was interested in atoms and just made some considerations.

Complexity level	Cognitive Process	Core aspect
Level II	Organizing	NOSI-4

Item AM5

Which typical feature for science is represented by Thomson's and Rutherford's quest for a model of the atom? Complete the sentence.

Scientific knowledge…

☑   … can change.

☐   … is true.

☐   … is random.

☐   … is hidden in nature.

Complexity level	Cognitive Process	Core aspect
Level V	Reproducing	NOS-3

Item AM6

This item is part of the item pool of the ESNaS project and therefore has to remain undisclosed.

Complexity level	Cognitive Process	Core aspect
Level III	Selecting	NOSI-5

Boyle's Law

Sources: Harré, 1984; Hermann, 1972; Neville, 1962; Webster, 1965

Stem 1

Robert Boyle (1627-1691) was a scientist and was interested in the properties of air and other gases. He conducted an experiment using a u-pipe filled with mercury (see picture 1). In the closed part (left side of the tube), a space filled with air formed. When he added more mercury, the space filled with air became smaller (see picture 2). Boyle recorded a series of measurement, in which he contrasted the altitude of mercury in the open part of the tube (right side of the tube) with the size of the space with air in the closed part. These measurements resulted in that what we know as law $p \cdot V = const$, today. Though, Boyle just published only his data and not the law, first. He did not realize that these data could be expressed in terms of a law. It was only when Richard Towneley, another researcher, interpreted the data using this law. Towneley talked about it with Boyle. Boyle then took Towneley's suggested interpretation and, in a second article, he published his data together with the law. At that time, Boyle called the law "Towneley's hypothesis", today, we know it as "Boyle's law".

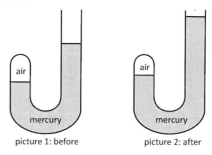

picture 1: before   picture 2: after

Boyle's experimental setup

Item BL1

This item is part of the item pool of the ESNaS project and therefore has to remain undisclosed.

Complexity level	*Cognitive Process*	*Core aspect*
Level I	Selecting	NOSI-6

Item BL2

This item is part of the item pool of the ESNaS project and therefore has to remain undisclosed.

Complexity level	Cognitive Process	Core aspect
Level V	Integrating	NOSI-6

Item BL3

What did Boyle have to do in order to formulate his law?

☐   He published the data which in itself are the law.

☐   He rejected the data because they did not show the law he wanted to find.

☑   He interpreted the data following Towneley's suggestion.

☐   He asked Towneley for help as data collected by two researchers are truer.

Complexity level	Cognitive Process	Core aspect
Level III	Organizing	NOS-2

Item BL4

Was Towneley acting scientifically when he only used the formula to interpret Boyle's data?

☐   No. Towneley would have had to test an hypothesis.

☐   No. Towneley would have had to modify at least one variable.

☐   Yes. Towneley was following the scientific method.

☑   Yes. Towneley applied mathematics to the unchanged data.

Complexity level	Cognitive Process	Core aspect
Level III	Organizing	NOSI-5

Item BL5

How did Towneley contribute to scientific progress? Complete the sentence.

Using the formula $p \cdot V = const$, Towneley interpreted…

☐   … an hypothesis.

☐   … an experiment.

☐   … a theory.

☑   … a set of data.

Complexity level	Cognitive Process	Core aspect
Level I	Reproducing	NOSI-6

Stem 2

Scientific knowledge is not produced automatically by thinking and experimenting. There are a lot of different influences, for example, a researcher's character, contact with other researchers, new inventions, and so on.

Robert Boyle (1627-1691) was a British scientist and was interested in the properties of air and other gases. He learned about a German invention, the air pump. This pump could be used to produce a vacuum. Boyle was thrilled about this invention and, thus, wanted to produce a vacuum on his own, too. Therefore, he developed a new air pump. Boyle's work on the properties of air and gases was influenced by his new invention. Using this pump he conducted numerous experiments with air.

Item BL6

How was Boyle's work influenced?

☐   Boyle's creativity did not influence his work on air and gases.

☐   The German air pump slowed down his work on air and gases.

☑   The quest for developing an air pump encouraged his work on air and gases.

☐   Findings of other researchers did not affect his work on air and gases.

Complexity level	*Cognitive Process*	*Core aspect*
Level III	Reproducing	NOS-1

Item BL7

This item is part of the item pool of the ESNaS project and therefore has to remain undisclosed.

Complexity level	*Cognitive Process*	*Core aspect*
Level I	Reproducing	NOSI-4

Item BL9

In his experiments, Boyle could show that air can be compressed and expanded. Boyle also strived to express this observation in a law. This caused him to conduct further experiments. Within one of these experiments he finally recorded data that represented what we know as "Boyle's law" $p \cdot V = const$, today. However, Boyle did not find out this formula on his own, but only by having discussions with another researcher named Richard Towneley.

Boyle generated new scientific knowledge. How was this knowledge influenced? Complete the sentence.

Scientific knowledge is…

☐ … fully objective. Influencing factors like, for example, the researcher's personality are excluded, so there is always agreement on findings.

☐ … true by itself. Therefore, it cannot be affected by factors like the researcher's conditions of living and work.

☐ … gained by conclusions from data. When concluding findings from data, the researcher's creativity is unimportant.

☑ … produced by human beings. Therefore, it is influenced, for example, by the researcher's personality or by his or her discourse with other researchers.

Complexity level	Cognitive Process	Core aspect
Level V	Reproducing	NOS-1

The Energy Concept

Sources: Hermann, 1972; Hund, 1972; Kleinert, 1980; Schirra, 1991;Schreier, 1984; Schreier, 2002; Simonyi, 2004

Stem 1

The concept 'energy' has not always had the meaning it has today. At the end of the 18th century, scientists agreed on the term 'mechanical energy.' However, it was not known that several forms of energy exist and that one form of energy can be transformed into another form.

There were two researchers, Julius Mayer and James Prescott Joule, essentially contributing to broaden the conception of energy. Mayer deduced from theoretical considerations that mechanical work can be transformed into thermal energy. Moreover, he was convinced that the sum of all energies would have to be constant during all processes. Independently from Mayer's work, Joule conducted a series of experiments from which he concluded the same concept that Mayer had found theoretically before. Through Mayer's and Joule's work the concept of energy fundamentally changed. 'Energy' is an important entity, which still is of central relevance in physics today.

Item en1

This item is part of the item pool of the ESNaS project and therefore has to remain undisclosed.

Complexity level	Cognitive Process	Core aspect
Level IV	Reproducing	NOS-3

Item en2

There were two different methods of research that influenced the development of the energy concept. Which?

☑ Mayer worked theoretically and Joule conducted an experiment.

☐ Joule collected data and Mayer interpreted them.

☐ _ Mayer stated a theory and Joule disproved it.

☐ Joule proposed hypotheses and Mayer tested them theoretically.

Complexity level	Cognitive Process	Core aspect
Level II	Reproducing	NOS-3

Item en3

Mayer and Joule approached the question of clarifying the concept of energy differently. Did both of them work scientifically?

☐ No. Only one of them worked scientifically. Scientists always have to vary variables in their investigations, which cannot be done if their work is just theoretical.

☐ Yes. In science, the only way for gaining knowledge is to conduct experiments. Theoretical work is only scientific when it fits experiments.

☑ Yes. In science, there is not only one method of research. Scientists can work differently: they work theoretically, conduct experiments or make observations.

☐ No. Only one of them worked scientifically. Scientists always have to propose hypotheses that have to be tested by experimental investigations.

Complexity level	Cognitive Process	Core aspect
Level V	Integrating	NOSI-5

Item en4

Both, Mayer and Joule, scientifically approached the advancement of the energy concept. What was Mayer's approach?

☐ He tested hypotheses with data.

☐ He conducted experiments.

☑ He made theoretical assumptions.

☐ He recorded a series of measurement.

Complexity level	Cognitive Process	Core aspect
Level I	Selecting	NOSI-5

Stem 2

It has not always been known that several forms of energy exist and that the sum of all energies in one system is conserved. At the end of the 18th century, scientists had an idea of mechanical work and mechanical energy. Some years later, Julius Robert Mayer wondered about combustion processes and heat production of living organisms. Working on that topic he had to accurately think about the heat concept. Mayer liked to work theoretically and by theoretical thinking, he finally developed a concept of 'energy' that also included heat and the conservation of energy. Some years later, another physicist, James Prescott Joule, worked on steam engines and heat effect of electric current. This work confirmed his conviction that there have to be other forms of energy than just the mechanical one. He finally conducted a number of experiments that supported the concept Mayer had developed before. Thus, the concept of 'energy' changed through Mayer's and Joule's work.

Item en5

Which statement about the concept of 'energy' is correct? Complete the sentence.

The concept of 'energy'…

☑ … was influenced by Mayer's interest in the heat production of organisms and by Joule's interest in steam engines.

☐ … has always been in nature and Mayer's and Joule's personal opinions and perceptions of energy prevented finding it.

☐ … had to be found, but its discovery was delayed by Mayer's and Joule's non-physical interest in steam engines and organisms.

☐ … was given by data that were only reliable by Mayer's and Joule's suppressing all subjective factors during their work.

Complexity level	Cognitive Process	Core aspect
Level IV	Reproducing	NOS-1

Item en6

Mayer and Joule worked on advancing the concept of energy. Which terms best describe their work?

☐   Mayer: experiment; Joule: theory

☐   Mayer: hypothesis; Joule: observation

☑   Mayer: theoretical thinking; Joule: experiment

☐   Mayer: law; Joule: hypothesis

Complexity level	Cognitive Process	Core aspect
Level II	Reproducing	NOS-2

Item en7

This item is part of the item pool of the ESNaS project and therefore has to remain undisclosed.

Complexity level	Cognitive Process	Core aspect
Level V	Integrating	NOS-1

Item en8

Mayer and Joule worked differently. What can be said about their work?

☐   Mayer and Joule had different questions they were working on. Working scientifically, they would have come to different results.

☐   Mayer and Joule discovered the same principle. Thus, they have worked on the same research question with the same methods.

☑   Mayer and Joule started investigations with different questions. Each question guided their individual research.

☐   Mayer and Joule worked on the same question. Since the question did not influence their methods, they approached it differently.

Complexity level	Cognitive Process	Core aspect
Level IV	Organizing	NOSI-4

Item en9

This item is part of the item pool of the ESNaS project and therefore has to remain undisclosed.

Complexity level	Cognitive Process	Core aspect
Level II	Reproducing	NOSI-4

The Universe

Sources: de Boer, 2007; Kinnebrock, 2002; Sharov, & Novikov, 1994; Simonyi, 2004

Stem 1

In the early 20th century, scientists thought that the whole universe was made up of only one galaxy, which also included the Earth. There also was an estimation of this galaxy's dimension. Beside stars and planets, diffuse light spots could be observed in outer space. Scientists thought that these light spots were nebulae of gas and dust that were part of our galaxy. Edwin Hubble observed some of these nebulae and calculated their distance from the earth, based on his observations. In doing so, he found out that some nebulae were quite far away from the earth so that they could not be situated in our galaxy. This led to the view that the universe contains more galaxies than just one.

Item gal1

How did Hubble's work further the progress of scientific knowledge?

☐   Hubble made observations on nebulae in outer space. These observations were disproved.

☐   Hubble made observations on nebulae in outer space. These observations will always be duplicated.

☐   Scientists thought that the Earth's galaxy was the whole universe. Hubble's results proved this view.

☑   Scientists thought that the Earth's galaxy was the whole universe. Hubble's results changed this view.

Complexity level	*Cognitive Process*	*Core aspect*
Level II	Selecting	NOS-3

Item gal2

How did Hubble's work further the progress of scientific knowledge?

☐   The estimation of our galaxy's dimensions was proved right by Hubble's observations and calculations.

☑   The prevalent view of the universe changed due to Hubble's observations and calculations.

☐   Hubble's new observations were seen as measurement errors due to the prevalent view of the universe.

☐   A new model of the universe was rejected due to the estimation of the galaxy's dimensions.

Complexity level	*Cognitive Process*	*Core aspect*
Level III	Selecting	NOS-3

Item gal3

What happened when diffuse light spots were observed in outer space?

☐ Hubble's theory of other galaxies was wrong. His calculations contained too many measurement errors.

☐ Hubble showed that all of them were nebulae. The estimation of our galaxy's dimensions was proved correct.

☑ Scientists considered them to be nebulae. Hubble showed that some of them were other galaxies.

☐ Scientists showed that they were part of our galaxy. Some of them were seen as measurement errors.

Complexity level	*Cognitive Process*	*Core aspect*
Level II	Selecting	NOSI-6

Item gal4

Being a scientist, what did Hubble do in his research?

☐ Hubble observed different nebulae and showed that the universe was made up of one single galaxy.

☑ Hubble observed different diffuse light spots and calculated their distance from the earth.

☐ Hubble investigated nebulae in outer space and modified their distance from the earth systematically.

☐ Hubble investigated diffuse light spots and showed that some of them were measurement errors.

Complexity level	*Cognitive Process*	*Core aspect*
Level II	Selecting	NOSI-5

Stem 2

In the early 20th century, scientists wondered about the universe. However, even though they investigated the same universe, they achieved different results. For example, Albert Einstein and Willem de Sitter thought that the universe was static and rested in itself. Many scientists thought the same. Particular workings which assumed a dynamic universe were ignored. In 1927, Georges Lemaître found a theoretical error in de Sitter's model. Therefore, he assumed another model of the universe that it was expanding and getting larger over time.

Item gal5

Using Lemaître's model, some observations about neighboring galaxies could be explained.

Which statement about the views on the universe in the early 20th century is correct?

☑ Lemaître developed a new model of the universe by which observations could be explained. Thus, the view of the universe changed.

☐ Many scientists believed in a static universe as proposed by Einstein and de Sitter. Since everyone agreed on this, this view will always be correct.

☐ Because Lemaître's model could be used for explaining observations, it was proved right. Thus, new observations cannot lead to improving this view.

☐ Two different models were respectively suggested by Einstein and de Sitter describing the universe. Since there is only one universe, one of them must be wrong.

Complexity level	Cognitive Process	Core aspect
Level IV	Selecting	NOS-3

Item gal6

What did Einstein, de Sitter, and Lemaître do when they did research on the universe?

☑ They observed the same universe and they got different results.

☐ They observed nature which can be described in only one way.

☐ They invented theories which are seen as their individual opinions.

☐ They invented laws and proved which the correct one was.

Complexity level	Cognitive Process	Core aspect
Level II	Reproducing	NOSI-6

Stem 3

In science, there is not only one way to get results. Scientists conduct experiments and modify variables in order to collect data, but they also make observations, for example. Moreover, there are scientists that develop theories in order to explain observations and measurement results.

In the early 20th century, there were different scientific models describing the universe. Albert Einstein and Willem de Sitter thought of a universe that was static and rested in itself. Georges Lemaître, thought of a universe that was expanding and getting larger over time. From this model, Lemaître derived a law on the motion of neighboring galaxies. Independently, the physicist Edwin Hubble made observations and calculations from which he derived the same law. Combing both of their work resulted in a view on the origin and the development of the universe. Using it, Hubbles observations could be explained and it is still accepted, today.

Item gal7

What does Hubble's and Lemaître's work show about science in general?

☐   In science, there is only one method to follow in order to get results.

☐   Scientists have to work through experimental instructions to find the right results.

☐   In science, theories do not generate knowledge as precisely as experiments do.

☑   Scientists use different and various approaches in order to gain new knowledge.

Complexity level	*Cognitive Process*	*Core aspect*
Level V	Reproducing	NOSI-5

Item gal8

Which of the following statements best describes Hubble's research on the universe?

☐   Hubble investigated the universe and explained a theory.

☑   Hubble made observations and inferred a law.

☐   Hubble developed a law and it was proved correct.

☐   Hubble collected data and the data were a new law.

Complexity level	*Cognitive Process*	*Core aspect*
Level II	Selecting	NOS-2

Stem 4

In the early 20th century, there were different scientific models describing the universe. Albert Einstein and Willem de Sitter thought of a universe that was static and rested in itself. Georges Lemaître, thought of a universe that was expanding and getting larger over time. From this model, Lemaître derived a law on the motion of neighboring galaxies. Independently, the physicist Edwin Hubble made observations and calculations from which he derived the same law. Combing both of their work resulted in a view on the origin and the development of the universe. Using it, Hubbles observations could be explained and it is still accepted, today.

Item gal9

In what way did Hubble's and/or Lemaître's work further the progress in scientific knowledge?

☐ Because in science only data lead to new knowledge, only Hubble's work contributed to the progress of science. Even if Lemaître discovered the same law, his work was not important.

☐ Neither Lemaître's work nor Hubble's contributed to the progress of science, because there already were two models many scientists believed in. In science, laws and observations have to fit well established theories.

☐ Because in science only theories can explain observations, only Lemaître's work contributed to the progress of science. Hubble's observations were not necessary for supporting Lemaître's work.

☑ Lemaître's theory was strengthened by Hubble's observations and calculations. In science, theories are needed to explain observations, and they have to be supported by data.

Complexity level	*Cognitive Process*	*Core aspect*
Level IV	Integrating	NOS-2

Item gal10

Einstein, de Sitter, and Lemaître did research on the same universe. How could they develop three different models of the universe?

☐ Scientific models represent nature identically. Only reliable data lead to steady models.

☐ In science, only one model describes nature correctly and others had too many errors.

☐ Scientific models are individual opinions. Only laws can describe nature how it really is.

☑ In science, models can be inferred from data or derived from theories in various ways.

Complexity level	*Cognitive Process*	*Core aspect*
Level V	Organizing	NOSI-6

Item gal11

This item is part of the item pool of the ESNaS project and therefore has to remain undisclosed.

Complexity level	Cognitive Process	Core aspect
Level V	Integrating	NOSI-6

Faraday's Experiment on Induction

Sources: Fraunberger, 1985; Heller, 1965; Hermann, 1972; Kleinert, 1980; Schreier, 1984; Schreier, 2002; Simonyi, 2004

Stem 1

Scientists always work on a specific topic of research and a particular research question. They orientate their scientific investigation to these questions.

Michael Faraday lived in the 19th century and worked with a chemist. At that time, the scientist Hans Christian Oersted had conducted an experiment that showed that electricity could change a magnetic needle's direction. Scientists wondered if magnetism could affect electricity, too. Faraday learned about Oersted's experiment and became interested in electromagnetism. He also wondered if magnetism could affect electricity. He conducted numerous experiments and many of them had no successful outcome. In 1831, he finally did a decisive experiment. His observations allowed the conclusion and, hence, the answer to his question: changing a magnetic field causes an electric current.

Item ind1

Why did Faraday conduct physics experiments?

☐ Faraday wanted to work free from subjective influences.

☑ Faraday was curious about electromagnetism.

☐ Faraday was really looking for non-creative research.

☐ Faraday tried to disprove Oersted's findings.

Complexity level	Cognitive Process	Core aspect
Level I	Reproducing	NOS-1

Item ind2

Which role did Faraday's research question play in his research? Compare it to scientific research in general.

☐   Research questions have to be formulated as hypotheses. If Faraday had stated an hypothesis, his experiments would have been more successful.

☑   Scientific research is oriented toward research questions. Faraday's investigations were led by his wondering if magnetism could affect electric current.

☐   Scientific research is slowed down by research questions. If Faraday had just been doing research without a question, he would have found his results more quickly.

☐   Research questions are only meaningful if a specialized scientist works on it. Since Faraday's question was closely related to Oersted's, Oersted would have to answer it.

Complexity level	*Cognitive Process*	*Core aspect*
Level V	Organizing	NOSI-4

Item ind3

How did Faraday's personal character influence his research on electromagnetism?

☐   Faraday's creativity had to be suppressed in order to make the correct observations.

☐   Faraday's inventive thoughts were the reason for a number of his experiments not having successful outcomes.

☐   Faraday was distracted from his own research because of having heard from Oersted's work.

☑   Faraday did a crucial experiment because he was interested in electromagnetism.

Complexity level	*Cognitive Process*	*Core aspect*
Level III	Selecting	NOS-1

Item ind4

What role do research questions play in scientific investigations like Faraday's?

☑   Scientific investigations are oriented towards research questions.

☐   Research questions have to be formulated as hypotheses.

☐   The direct way to scientific knowledge is disguised by research questions.

☐   Research questions can only be answered by one single scientist.

Complexity level	*Cognitive Process*	*Core aspect*
Level V	Reproducing	NOSI-4

Stem 2

Michael Faraday lived in the 19th century and worked with a chemist. At that time, the scientist Hans Christian Oersted had conducted an experiment that showed that electricity could change a magnetic needle's direction. Scientists wondered if magnetism could affect electricity, too. Faraday learned about Oersted's experiment and became interested in electromagnetism. He also wondered if magnetism could affect electricity. He conducted numerous experiments and many of them had no successful outcome. In 1831, he finally did a decisive experiment. His observations allowed the conclusion and, hence, the answer to his question: changing a magnetic field causes an electric current.

Item ind5

How important was Faraday's research question to his research and findings?

☑ Faraday conducted experiments because he wondered if the relationship found by Oersted was valid if reversed.

☐ It took Faraday a long time to get to his findings because he had not stated a pure hypothesis.

☐ Faraday's research question was meaningless because the relationship he found was already seen in nature.

☐ The research question disguised Faraday's findings because, actually, it was part of Oersted's field of research.

Complexity level	*Cognitive Process*	*Core aspect*
Level III	Reproducing	NOSI-4

Item ind6

What would other scientists necessarily have to do if they were in Faraday's place? Complete the sentence.

They would have to …

☐ … hold back any inference because it covers the true results.

☐ … study the data closely because they are the real findings.

☑ … interpret the observations in order to get the results

☐ … simply observe nature because findings emerge from it.

Complexity level	*Cognitive Process*	*Core aspect*
Level III	Integrating	NOS-2

Item ind7		
This item is part of the item pool of the ESNaS project and therefore has to remain undisclosed.		
Complexity level	*Cognitive Process*	*Core aspect*
Level II	Selecting	NOS-1

Stem 3

Michael Faraday lived in the 19th century and worked with a chemist. At that time, the scientist Hans Christian Oersted had conducted an experiment that showed that electricity could change a magnetic needle's direction. Scientists wondered if magnetism could affect electricity, too. Faraday learned about Oersted's experiment and became interested in electromagnetism. He also wondered if magnetism could affect electricity. At first, he repeated Oersted's experiment and extended it. He then conducted numerous experiments and many of them had no successful outcome. In 1831, he finally did a decisive experiment. His observations allowed the conclusion and, hence, the answer to his question: changing a magnetic field causes an electric current.

Item ind8		
How did Faraday find out that changing a magnetic field causes an electric current?		
☑ Faraday copied his research question from Oersted's work. Since Faraday did not create and work on a research question he invented on his own, his findings were not reliable.		
☐ Faraday's research question caused a lot of errors during his experiments. This would not have had been happened if he had stated hypotheses instead of a question.		
☐ Faraday's research question interfered with the research of a second scientist. The findings could only be discovered because two scientists worked in the same field.		
☐ Oersted's findings inspired Faraday to wonder if, in reverse, magnetism could affect an electric current, too. Faraday's investigations were oriented towards this question and finally led to his findings.		
Complexity level	*Cognitive Process*	*Core aspect*
Level IV	Reproducing	NOSI-4

Item ind9		
This item is part of the item pool of the ESNaS project and therefore has to remain undisclosed.		
Complexity level	*Cognitive Process*	*Core aspect*
Level I	Reproducing	NOSI-5

Item ind10

What happened during Faraday's investigations? Compare it to today's research.

☑ Faraday conducted various experiments and many of them failed. Today's research still does not proceed along straight lines without errors because there is no exclusive method to follow in science.

☐ Faraday wanted to answer only one research question. Today's research is much broader in scope and, therefore, it generates new knowledge more purposefully.

☐ Faraday did not follow an accurate guideline. Today's research is not slowed down by measurement errors because it has more accurate and elaborated research instructions available.

☐ It took Faraday a long time to get to his findings. Today's research proceeds only by stating and testing hypotheses and, thus, it is less time-consuming.

Complexity level	Cognitive Process	Core aspect
Level IV	Integrating	NOSI-5

The Composition of the Atomic Nucleus

Sources: Chadwick, 1932a; Chadwick, 1932b; Rutherford, 1920; Simonyi, 2004

Stem 1

Today, when scientists imagine atomic nuclei they think about protons and neutrons. This wasn't always the case. In 1911, the physicist Ernest Rutherford developed the theory that nuclei were only made up of small positive charged particles. He named them protons. Using this model he was able to explain observations he had made in experiments before.

Later (1913), Frederick Soddy discovered isotopes. Isotopes are atoms having the same number of protons but different masses as other atoms of the same element. When Rutherford learned of Soddy's work, he tried to explain this discovery using his own idea of the atomic nucleus. However, Rutherford realized that this discovery did not agree with his own idea about the nucleus. In order to be able to explain Soddy's discovery, Rutherford developed a new theory of the nucleus. He thought there had to be another type of particle in the nucleus besides protons.

Item neu1

This item is part of the item pool of the ESNaS project and therefore has to remain undisclosed.

Complexity level	Cognitive Process	Core aspect
Level III	Reproducing	NOS-1

Item neu2

As a scientist, was Rutherford correct in developing a new model?

☑   Yes. In science, models must fit observations already made.

☐   No. In science, discoveries can never lead to new models.

☐   Yes. A scientist can change his/her model without taking other results into account.

☐   No. The scientist that made the discovery is the only one that can change the model.

Complexity level	Cognitive Process	Core aspect
Level V	Integrating	NOS-2

Item neu3

What is the best description of Rutherford's and Soddy's work?

☐   Soddy suggested a law and Rutherford tried to prove it.

☑   Soddy made a discovery and Rutherford tried to explain it.

☐   Rutherford had a theory and Soddy tried to verify it.

☐   Rutherford made an observation and Soddy tried to refute it.

Complexity level	Cognitive Process	Core aspect
Level II	Organizing	NOSI-6

Item neu4

Some years later, the physicist Chadwick succeeded in detecting the particles suggested by Rutherford.

What does this story show about scientific knowledge? Complete the sentence.

Scientific knowledge…

☑   … is not absolute or certain. It can change because of new observations or new theories.

☐   … is given through experimental data. Because the data are exact the knowledge is definite.

☐   … is based only on theories. Because these theories are personal opinions the knowledge is unreliable.

☐   … is inherent in nature. It is true because there is no creative tolerance when found experimentally.

Complexity level	Cognitive Process	Core aspect
Level V	Integrating	NOS-3

Item neu5

Rutherford's first model of the nucleus assumed protons as the only particles in the nucleus. Was this model scientific?

☐   Yes, because he devised it directly from nature.

☑   Yes, because using it he could explain observations.

☐   No, because it did not agree with Soddy's discovery.

☐   No, because being not complete it had to be changed.

Complexity level	Cognitive Process	Core aspect
Level IV	Selecting	NOS-2

Stem 2

Ernest Rutherford was a physicist. In 1920, he developed a theory of the atomic nucleus consisting of smaller particles called protons and neutrons. Protons could be detected but the neutrons had not been found yet. Rutherford discussed his theory with his student James Chadwick and they worked together in order to detect the neutrons. However, they didn't succeed.

Some years later, Walther Bothe and Hans Geiger conducted other experiments. In doing so, they observed rays with high energy but they did not know what these rays were. While Chadwick was still searching for the neutron, he learned about Bothe and Geiger's experiments. He supposed that the rays perhaps could be neutrons. He conducted Bothe and Geiger's experiment himself and was able to show that the observed rays were fast-moving, high-energy particles. Moreover, he could confirm that these particles were the same that Rutherford had predicted.

Item neu6

What occurred when Chadwick repeated Bothe and Geiger's experiment?

☑   Chadwick confirmed Rutherford's theory.

☐   Chadwick remained unconvinced of Rutherford's theory.

☐   Chadwick was not able to explain the outcome.

☐   Chadwick changed the experiment.

Complexity level	Cognitive Process	Core aspect
Level I	Reproducing	NOS-2

Item neu7

When replicating Bothe and Geiger's experiment, how did Chadwick further the discovery of the neutrons?

☑   He interpreted the outcome of the known experiments differently from other researchers.

☐   He used improved devices by which neutrons could be observed more accurately.

☐   He was the first who produced neutrons during an experiment.

☐   He collected his own data in which the neutron's discovery was included.

Complexity level	Cognitive Process	Core aspect
Level III	Organizing	NOSI-6

Item neu8

What encouraged the discovery of the neutron?

☐   Bothe and Geiger's theory of neutrons and Chadwick's proof of this theory

☐   Bothe and Geiger's interpretation and Chadwick's new devices

☑   Bothe and Geiger's findings and Chadwick's conviction of the neutron's existence

☐   Bothe's and Geiger's detection of the neutron and Chadwick's theory

Complexity level	Cognitive Process	Core aspect
Level II	Organizing	NOS-1

Item neu9

Why did Chadwick repeat Bothe and Geiger's experiment? Complete the sentence.

Chadwick wanted to …

☑   … search for the neutron.

☐   … develop a theory of nuclei.

☐   … prove Bothe and Geiger's hypothesis.

☐   …improve Bothe and Geiger's investigation.

Complexity level	Cognitive Process	Core aspect
Level I	Selecting	NOSI-4

Newton and Colored Light

Sources: Achilles, 1996; Nawrath, 2007

Stem

Isaac Newton is well known for his work on mechanics, but he also worked in the field of optics. When Newton studied at Cambridge, he helped his professor to prepare the edition of a textbook on optics. This work inspired him to work experimentally in this field. In 1672, Newton published a new theory about light and colors. In this article he claimed that white light consists of colored light. In addition, he reported about an experiment which supported his theory (picture 1 roughly shows the paths of some colored light rays): white light, at first, was dispersed by a prism into a colored spectrum, and a lens fused the light of different colors into white light again. Newton's theory raised a lot of discussions among other scientists. For instance, René Descartes had another theory on the nature of light and interpreted Newton's experiment differently. He thought that the colors were generated inside the prism; meaning that the white light became colored by the prism.

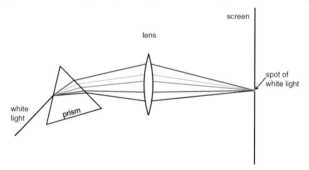

picture 1: Scheme of Newton's experiment on the colors of light

---

Item New1

Compare Newton's work with today's research. How is both influenced by scientists' personalities? Complete the sentence.

Scientists work…

☑   … is influenced by individual factors like his or her enthusiasm.

☐   … is hindered by individual attitudes, like creativity.

☐   … is free of opinions, like scientific knowledge is.

☐   … does not need any imagination, because knowledge is gained from data directly.

Complexity level	*Cognitive Process*	*Core aspect*
Level III	Integrating	NOS-1

Item New2

This item is part of the item pool of the ESNaS project and therefore has to remain undisclosed.

Complexity level	*Cognitive Process*	*Core aspect*
Level IV	Reproducing	NOS-2

Item New3

Why could Descartes get another theory of light even if he read the experimentally observed data reported by Newton?

☐   Descartes would have cooperated with Newton to get the same theory.

☐   Descartes made a logical mistake when he interpreted Newton's data.

☐   Descartes did not understand the data correctly when he read them.

☑   Descartes interpreted the data using his own theory of light.

Complexity level	*Cognitive Process*	*Core aspect*
Level III	Reproducing	NOSI-6

Nuclear Fission

Sources: Fermi, 1934a; Fermi, 1934b; Hermann, 1972; Noddack, 1934a; Noddack, 1934b; Simonyi, 2004

Stem

At the beginning of the 20th century, the heaviest element known was uranium. Later, many scientists aimed at producing new elements heavier than uranium. Enrico Fermi, for instance, bombarded uranium with neutrons (small non-charged particles). From his observations he inferred that a new, heavier element was generated. The chemist Ida Noddack read about Fermi's work. She suggested that it also could be possible that no new, heavier element was produced, but instead, the nuclei were split. Her considerations were not taken seriously by other scientists because they could not imagine that nuclear fission is possible. It was only when another research group repeated Fermi's experiment and analyzed the products more accurately, that nuclear fission was confirmed. After a theoretical explanation for this process was given, too, the scientific community accepted the fact of nuclear fission.

Item nufi1

How did it happen that Fermi stated the generation of a new element?

☑   Fermi interpreted his own observations.

☐   This result was Fermi's data itself.

☐   Others repeated Fermi's experiment.

☐   Others read about Fermi's findings.

Complexity level	*Cognitive Process*	*Core aspect*
Level I	Selecting	NOSI-6

Item nufi2

This item is part of the item pool of the ESNaS project and therefore has to remain undisclosed.

Complexity level	*Cognitive Process*	*Core aspect*
Level IV	Integrating	NOSI-4

Item nufi3

How could Fermi and Noddack come to different conclusions even if both of them referred to the same experiment?

☐   Fermi inferred the generation of a new element from his own observations. Noddack came to another finding because she did not conduct the experiment herself.

☐   Fermi made his observations from a physicist's point of view. Because Noddack was working in another field of research, she suggested a wrong conclusion.

☑   Fermi analyzed his observations not thinking of a possible nuclear fission. Noddack interpreted Fermi's results considering nuclear fission being possible.

☐   Fermi was shown the result directly from the data. Since Noddack did not have Fermi's raw data available to her, she interpreted his publication differently.

Complexity level	*Cognitive Process*	*Core aspect*
Level IV	Reproducing	NOSI-6

Item nufi4

During the discovery of nuclear fission, how did Fermi and Noddack contribute to scientific progress?

☑   Fermi conducted experiments. Noddack interpreted the outcomes theoretically.

☐   Fermi analyzed other scientists' data. Noddack made a discovery.

☐   Fermi developed a theory. Noddack repeated another scientist's experiments.

☐   Fermi suggested an hypothesis. Noddack came up with a model theoretically.

Complexity level	Cognitive Process	Core aspect
Level II	Organizing	NOSI-5

Item nufi5

This item is part of the item pool of the ESNaS project and therefore has to remain undisclosed.

Complexity level	Cognitive Process	Core aspect
Level IV	Integrating	NOSI-6

Item nufi6

First, scientists did not know that nuclear fission is possible, and finally they did. Which typical feature of scientific knowledge can be seen from that? Complete the sentence.

Scientific knowledge…

☐   … is fixed.

☐   … are the data itself.

☐   … has to be discovered.

☑   … can change.

Complexity level	Cognitive Process	Core aspect
Level V	Integrating	NOS-3

The Discovery of Electromagnetism

> **Sources:** Achilles, 1996; Heller, 1965; Hermann, 1972; Kleinert, 1980; Mason, 1974; Rosenberger, 1965; Schreier, 1984; Schreier, 2002; Simonyi, 2004

Stem 1

For a long time scientists thought that electricity and magnetism were distinct concepts. There were many observations related to only one of the two fields but none connecting both of them. In the beginning 19th century, Hans Christian Oersted was convinced that there had to be a connection between electricity and magnetism, like other scientists did, too. Therefore, he tried to find such a connection experimentally. Finally, Oersted succeeded in showing that a current flowing through a wire changes the direction of a magnetic needle and that the current's direction determines in which direction the needle rotates.

Item Oer1

This item is part of the item pool of the ESNaS project and therefore has to remain undisclosed.

Complexity level	*Cognitive Process*	*Core aspect*
Level III	Organizing	NOS-3

Item Oer2

Oersted's experiment changed conceptions about electricity and magnetism of those researchers who did not belief in a connection between the two concepts. Since then, many scientists have studied electromagnetism.

Why did scientists first study electricity and magnetism separately, and, later, electromagnetism instead? Complete the sentence.

The scientific concepts about electricity and magnetism…

☑   … changed.

☐   … became a law.

☐   … were hypotheses.

☐   … were proved.

Complexity level	*Cognitive Process*	*Core aspect*
Level I	Reproducing	NOS-3

Stem 2

In the 18th century, it was known that lightning could affect a magnetic compass needle. The scientist Hans Christian Oersted knew about this observation and investigated electricity and magnetism. Moreover, he was totally convinced that nature is systematic and unified. Therefore, he thought that electricity and magnetism were related to each other. In order to support his view of a unified nature, he did a lot of experiments to find a connection between electricity and magnetism. Finally, in 1819, he could show experimentally that an electric current changes the direction of a magnetic needle.

Item Oer3

Why did Oersted conduct a lot of experiments on electricity and magnetism?

☐ Loving experimental work he just did some research on electricity and magnetism without any purpose.

☑ According to his view of electricity and magnetism he wanted to support his idea of their relationship.

☐ Since testing hypotheses is the only way to achieve new knowledge, he wanted to test his idea with experiments.

☐ Since he did not believed in a compass needle being influenced by lightning, he conducted a lot of checking experiments.

Complexity level	Cognitive Process	Core aspect
Level III	Selecting	NOSI-4

Item Oer4

How was Oersted's research influenced? Complete the sentence.

Oersted's conducting a decisive experiment, was mainly influenced by...

☐ ... the theory developed before and his own objectivity.

☐ ... the law he discovered and the wrong ideas of other researchers.

☑ ... the observations done before and his own view of nature.

☐ ... the question he had and the true knowledge already known.

Complexity level	Cognitive Process	Core aspect
Level II	Reproducing	NOS-1

Ohm's Law

Sources: Achilles, 1996; Fraunberger, 1985; Heering, 2001; Hermann, 1972; Scharf, 2004; Schreier, 1984; Schreier, 2002; Simonyi, 2004

Stem 1

Ohm's law is very famous. However, it took a long time for it to become commonly known in the 19th century. Actually, Georg Simon Ohm was a teacher of physics and mathematics. The school where he worked had a well-equipped laboratory, in which he conducted his experiments. He had a great interest in physics and he especially wanted to determine the strength of electric current as a function of the length and size of a wire.

Item OL1

This item is part of the item pool of the ESNaS project and therefore has to remain undisclosed.

Complexity level	*Cognitive Process*	*Core aspect*
Level I	Reproducing	NOS-1

Item OL8

This item is part of the item pool of the ESNaS project and therefore has to remain undisclosed.

Complexity level	*Cognitive Process*	*Core aspect*
Level V	Reproducing	NOS-2

Item OL3

How was "Ohm's law" discovered?

☑ Ohm had some evidence from his experimental data. He then described this evidence mathematically.

☐ Ohm had some evidence from his experimental data. These data were identical to the law.

☐ Ohm studied the work of another researcher. Using only their work he could obtain the law mathematically.

☐ Ohm studied the work of another researcher. He then conducted his published experiment and discovered the law.

Complexity level	*Cognitive Process*	*Core aspect*
Level IV	Organizing	NOS-1

Stem 2

Ohm's law is very famous. However, it took a long time for it to become commonly known in the 19th century. Georg Simon Ohm wanted to determine the strength of electric current as a function of the length and size of a wire. For that, Ohm initially used a voltage source called "galvanic trough apparatus". This was a voltage source being quite common at that time. Ohm experimentally found a non-linear relationship among current, voltage, and resistance, but he had great difficulties, because his measured data changed without observable reason. He published his findings in a journal. The journal's editor suggested to him a newly invented voltage source. This new source produced constant voltage, something which the trough apparatus could not. Ohm then repeated his experiments with this new instrument and discovered the linear relationship we know as "Ohm's law."

Item OL4

At first, Ohm published a non-linear law. Then he published a linear law. Why did the law change? Complete the sentence.

Scientific laws…

☐   … are provisional. This is the reason why laws never can be replicated exactly by experiments.

☐   … cannot change, because the nature described by the laws never changes.

☐   … are absolute. If there is another version of a law, this is an error of a scientist.

☑   … can change due to the design of improved instruments.

Complexity level	*Cognitive Process*	*Core aspect*
Level III	Integrating	NOS-3

Item OL9

Furthermore, Ohm composed a mathematic formula expressing his data appropriately. Ohm did not yet entirely comprehend the physical meaning of the formula's elements, though. However, his theoretical description was in line with his measured data. This is important, since in science, theories and laws always have to be in line with data. Finally, he published his measured data and the theoretically derived mathematical relationship we know as "Ohm's law", today.

What did Ohm have to do in order to make sure that he worked scientifically? Complete the following sentence.

Ohm made sure that his theoretical description was in line with…

☑   … the data he had collected.

☐   … the question he tried to answer.

☐   … the opinions of other researchers on that topic.

☐   … the ideas on that topic he had in mind.

Complexity level	Cognitive Process	Core aspect
Level I	Selecting	NOS-2

Stem 3

Ohm's law is very famous and important. However, it took a long time for it to become common-ly known in the 19th century. Georg Simon Ohm wanted to determine the strength of electric cur-rent as a function of the length and size of a wire. Ohm conducted a lot of measurements. Finally, he drew from his data a relationship among those entities we now know as current, voltage, and resistance. Furthermore, he composed a mathematic formula expressing his data appropriately. In doing so, a French scientist's theory on heat transport was quite helpful for him. Ohm did not yet entirely comprehend the physical meaning of the formula's elements, though. Finally, he pub-lished his measured data and the theoretically derived mathematical relationship we know as "Ohm's law", today. Nevertheless, Ohm's work was not accepted broadly, because a lot of scien-tists did not realize the meaning of a mathematical description.

Item OL5

This item is part of the item pool of the ESNaS project and therefore has to remain undisclosed.

Complexity level	Cognitive Process	Core aspect
Level III	Selecting	NOS-2

Stem 4

Ohm's law is very famous and important. However, it took a long time for it to become commonly known in the 19th century. Actually, Georg Simon Ohm was a teacher of physics and mathematics. The school where he worked had a well-equipped laboratory, in which he conducted his experiments. He had a great interest in physics and he especially wanted to determine the strength of electric current as a function of the length and size of a wire. In order to determine this relationship, Ohm used a voltage source and wires of various measures of length and size. Above the wire, he hung a compass needle on a string. When there was a current in the wire, the compass needle turned and the string twisted (see picture 1). Ohm assumed that the string torque was a good measure of the strength of electric flux.

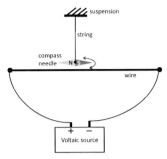

picture 1: scheme of Ohm's experimental set-up

Item OL6

Did Ohm work scientifically?

☐   Ohm worked scientifically because he was a scientist.

☐   Ohm did not work scientifically because he did not prove an hypothesis.

☑   Ohm worked scientifically because he tried to answer a question on nature.

☐   Ohm did not work scientifically because he did not measure the current with an ammeter.

Complexity level	*Cognitive Process*	*Core aspect*
Level III	Integrating	NOSI-4

Item OL7

For his experiments, Ohm initially used a voltage source called "galvanic trough apparatus". This was a voltage source being quite common at that time. Ohm discovered a non-linear relationship among current, voltage, and resistance, but he had great difficulties, because his measured data changed without observable reason. He published his findings in a journal. The journal's editor suggested to him a newly invented voltage source. This new source produced constant voltage, something that the trough apparatus could not. Ohm then repeated his experiments with this new instrument and discovered the linear relationship we know as "Ohm's law."

Why did Ohm's findings change?

☐   Ohm's first findings were corrected by another scientist.

☑   Ohm repeated the experiment with improved instruments; this lead to new results.

☐   Ohm's non-linear law was in contradiction to another scientist's theory.

☐   Ohm's first findings were a non-linear relationship; this was inconsistent to his theory.

Complexity level	Cognitive Process	Core aspect
Level I	Selecting	NOS-1

Item OL2

Ohm became very famous for the law he discovered. Which factors were very supportive?

☐   Ohm was a teacher of physics and tried to educate his pupils as well as possible.

☑   Ohm was interested in physics and he could work in a well-equipped laboratory.

☐   Ohm was interested in physics and he tried to educate his pupils as well as possible.

☐   Ohm worked in a well-equipped laboratory and he was instructed by school authority to do it.

Complexity level	Cognitive Process	Core aspect
Level II	Selecting	NOS-1

Superconductivity

Sources: Deger, Gleixner, Pippig, & Worg, 2001; Hermann, 1972; Jorda, 2008; Reif- Acherman, 2004

Stem 1

In science, knowledge is gained through the interplay of theoretical and experimental work.

At school we learn that every conductor has an electric resistance. However, in 1911, Heike Kamerlingh Onnes discovered superconductivity, which is a state in which metals show zero resistance. Some time before this discovery, Kamerlingh Onnes met the physicist Johannes Diderikvan der Waals and learned about his theory of gases. Using his theory, van der Waals had developed a formula describing the relationship among volume, pressure and temperature of a gas. Kamerlingh Onnes was determined to test the validity of van der Waals's theory for a large range of temperature. In order to achieve this, he invented some refrigeration devices, which made possible experiments at very low temperatures. These new devices made a whole new field of research possible. Kamerlingh Onnes could show that the formula of van der Waals was valid at low temperatures. Other questions could also be investigated, though. For instance, measuring the electric resistance of metals at very low temperatures was possible. At that time, there were many different ideas on what happens to a metal's resistance at very low temperatures. Kamerlingh Onnes investigated that question and found superconductivity.

Item SL1

This item is part of the item pool of the ESNaS project and therefore has to remain undisclosed.

Complexity level	*Cognitive Process*	*Core aspect*
Level IV	Selecting	NOSI-4

Item SL2

This item is part of the item pool of the ESNaS project and therefore has to remain undisclosed.

Complexity level	*Cognitive Process*	*Core aspect*
Level II	Organizing	NOSI-5

Item SL3

This item is part of the item pool of the ESNaS project and therefore has to remain undisclosed.

Complexity level	*Cognitive Process*	*Core aspect*
Level V	Organizing	NOS-2

Item SL4

This item is part of the item pool of the ESNaS project and therefore has to remain undisclosed.

Complexity level	*Cognitive Process*	*Core aspect*
Level I	Reproducing	NOS-3

Item SL5

What was the first step in Kamerlingh Onnes's discovery of superconductivity?

☐   He just was interested in conducting experiments.

☐   He stated the hypothesis that there is superconductivity.

☐   He invented refrigeration devices for no particular reason.

☑   He wanted to check van der Waals's formula at low temperature.

Complexity level	*Cognitive Process*	*Core aspect*
Level I	Selecting	NOSI-4

Item SL6

Which typical feature of scientific knowledge is shown by the discovery of superconductivity?

☐   New devices allowed investigations at low temperature and lead to the discovery of super-conductivity. This is an example that scientific knowledge is hidden in nature and has to be found.

☑   New devices allowed investigations at low temperature and lead to the discovery of super-conductivity. This is an example that scientific knowledge can change due to the invention of new devices.

☐   There were various theories about resistance at very low temperatures, but superconductivity could only be discovered by experiments. This is an example that scientific knowledge can be reworked only through empirical data.

☐   There were various theories about resistance at very low temperatures, but superconductivity could only be discovered by experiments. This is an example that scientific knowledge is certain only if it is found experimentally.

Complexity level	*Cognitive Process*	*Core aspect*
Level IV	Integrating	NOS-3

Item SL7

Which role did Kamerlingh Onnes's personal character play in discovering superconductivity?

☐ In order to discover superconductivity, he had to be convinced that there was superconductivity.

☐ In order to discover superconductivity, he had to hold back his creative thoughts.

☐ Becoming interested in the properties of gases delayed his research at low temperature.

☑ Becoming interested in the properties of gases led to his research at low temperature.

Complexity level	Cognitive Process	Core aspect
Level III	Organizing	NOS-1

Stem 2

At school we learn that every conductor has an electric resistance. However, in 1911, Heike Kamerlingh Onnes discovered superconductivity, which is a state in which metals show zero resistance. Some time before this discovery, Kamerlingh Onnes met the physicist Johannes Diderikvan der Waals and learned about his theory of gases. Using his theory, van der Waals had developed a formula describing the relationship among volume, pressure and temperature of a gas. Kamerlingh Onnes was determined to test the validity of van der Waals's theory for a large range of temperature. In order to achieve this, he invented some refrigeration devices, which made possible experiments at very low temperatures. These new devices made a whole new field of research possible. Kamerlingh Onnes could show that the formula of van der Waals was valid at low temperatures. Other questions could also be investigated, though. For instance, measuring the electric resistance of metals at very low temperatures was possible. At that time, there were many different ideas on what happens to a metal's resistance at very low temperatures. Kamerlingh Onnes investigated that question and found superconductivity.

Item SL8

This item is part of the item pool of the ESNaS project and therefore has to remain undisclosed.

Complexity level	Cognitive Process	Core aspect
Level I	Selecting	NOSI-5

Item SL9

Which two events caused the discovery of superconductivity?

☑ Kamerlingh Onnes wanted to prove van der Waals's formula on gases; other scientists raised the question what happens with the resistance at low temperature.

☐ Kamerlingh Onnes stated the hypothesis about superconductivity and other scientists required a proof for this hypothesis.

☐ Kamerlingh Onnes developed a formula for the relationship among volume, pressure and temperature of gases; other scientists wanted to verify it.

☐ Kamerlingh Onnes developed a theory on superconductivity and other scientists wondered if it was valid at low temperature.

Complexity level	Cognitive Process	Core aspect
Level II	Selecting	NOSI-4

Item SL10

How did Kamerlingh Onnes proceed on the way of discovering superconductivity?

☐ He heard about different ideas what happens with the resistance at very low temperature. By discovering superconductivity he contributed another idea.

☐ He was an enthusiastic inventor and became interested in research on resistance. To collect reliable data, he had to hold back his enthusiasm.

☑ He heard about van der Waals's formula and was decided to check it. To achieve this, he invented new devices.

☐ He was interested in the properties of gases. Because he actually wanted to discover superconductivity, this interest slowed down his quest.

Complexity level	Cognitive Process	Core aspect
Level IV	Selecting	NOS-1

Item SL11

How did Kamerlingh Onnes contribute to scientific progress?

☑ He conducted experiments in order to check a formula.

☐ He stated hypotheses in order to improve a model.

☐ He came up with a model in order to test an hypothesis.

☐ He developed a formula in order to collect data.

Complexity level	Cognitive Process	Core aspect
Level III	Reproducing	NOSI-5

Item SL12

When discovering superconductivity, Kamerlingh Onnes originally wondered about a theory on gases. What role do research questions play in scientific research, in the past and today?

☐   The direct way to scientific knowledge is covered by research questions.

☐   Research questions have to be formulated as hypotheses.

☑   Scientific investigations are oriented towards research questions.

☐   Research questions can only be answered by one scientist.

Complexity level	Cognitive Process	Core aspect
Level V	Integrating	NOSI-4

Item SL13

This item is part of the item pool of the ESNaS project and therefore has to remain undisclosed.

Complexity level	Cognitive Process	Core aspect
Level I	Selecting	NOS-3

The Solar System

Sources: Hakim, 2004; Hakim, 2005; Simonyi, 2004
Picture 1 by Tobias Viering according to Ron Miller in Hawking (1988, p. 15)

Stem

People have wondered about our solar system and what it looks like for a long time. In doing so, each model was dependent on their world view. In ancient times, Aristotle had the idea that the planets orbit the earth on different spheres (see picture 1).

Later on, orbits were observed which could not be explained with this model. Thus, Ptolemy developed a new model (see picture 2). However, as time went on, some new observations were made and they could not be explained even using Ptolemy's model. Based on this new information, scientists added more details to Ptolemy's model. Thus, it became very complex and confusing.

In late medieval times, Nicolaus Copernicus thought that there had to be a less complex model to describe the world. Thus, he worked out another model with the sun in the center instead of the earth (see picture 3). However, his model did not lead to more precise calculations than the old one. In some points, Johannes Kepler changed Copernicus' model and formulated three laws. He checked these laws with new, more precise observations. Using this advanced model the orbits of the planets could be predicted very precisely.

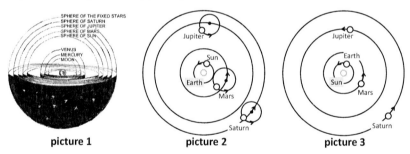

picture 1   **picture 2**   **picture 3**

Item SoSy1

All scientists observed the same solar system. In spite of this, why were there so many different models?

☐   Of all the models, only one describes the solar system correctly. All the others are individual opinions.

☑   New and more precise observations led to advanced models. Additionally, new models were suggested.

☐   Only one of the many different models is scientific. All the others are non-scientific assumptions.

☐   All the models described the same thing but are presented in different ways.

Complexity level	Cognitive Process	Core aspect
Level IV	Organizing	NOS-3

Item SoSy2

This item is part of the item pool of the ESNaS project and therefore has to remain undisclosed.

Complexity level	Cognitive Process	Core aspect
Level I	Selecting	NOS-2

Item SoSy3

Copernicus imagined a model with the sun in the center.

At that time, how could scientists decide if they were to trust Copernicus' model?

☐   They referred to books to see if someone had described the solar system this way before.

☐   They trusted the models that had been developed before and, thus, Copernicus' model had to be wrong.

☑   They observed the sky and checked if these observations fit Copernicus' model.

☐   They saw Copernicus' model as a theory, and, therefore, it was not important whether it is right or wrong.

Complexity level	Cognitive Process	Core aspect
Level IV	Organizing	NOS-2

Item SoSy4

Working as a scientist, how did Copernicus approach the question of the solar system's structure?

☐   He tested hypotheses and calculated formulae.

☐   He generated hypotheses and varied variables.

☑   He made observations and saw contradictions to the old model.

☐   He analyzed observations and conducted experiments.

Complexity level	Cognitive Process	Core aspect
Level II	Reproducing	NOSI-5

Item SoSy5

Copernicus wondered about the structure of the solar system. Was his approach to this question scientific?

☑   Yes. Making precise observations is scientific.

☐   Yes. Proving laws definitively right is scientific.

☐   No. To work scientifically he would have tested an hypothesis.

☐   No. To work scientifically he would have conducted experiments.

Complexity level	Cognitive Process	Core aspect
Level III	Integrating	NOSI-5

Item SoSy6

Aristotle already had modeled the solar system. Why did Ptolemy develop another model? Complete the sentence.

Aristotle's model...

☐   ... neglected laws that had been proved right before.

☐   ... was wrong because it had not become law.

☐   ... had been seen to be true for such a long time that it had to be improved.

☑   ... could only explain very few observations.

Complexity level	Cognitive Process	Core aspect
Level III	Reproducing	NOS-2

Item SoSy7

What was crucial for the development of a model of the solar system?

☐ New experiments were conducted and one theory could be proved right.

☐ Hypotheses were generated and tested by experiments.

☐ New devices were invented and the real model could be observed.

☑ Contradictions to observations came up and the explanatory model changed.

Complexity level	Cognitive Process	Core aspect
Level II	Organizing	NOS-3

Item SoSy9

In science, why were there so different models, even if they described the same the solar system? Complete the sentence.

All scientists ...

☐ ... thought about the solar system and only one of them discovered the true structure directly given by nature.

☐ ... had their own ideas of the solar system but only one of these was correct because it was given by observations.

☐ ... had to develop various theories because the observed data had been affected with measurement inaccuracies.

☑ ... observed the same solar system but interpreted the observations differently and, therefore, developed different models.

Complexity level	Cognitive Process	Core aspect
Level IV	Organizing	NOSI-6

Item SoSy11

This item is part of the item pool of the ESNaS project and therefore has to remain undisclosed.

Complexity level	Cognitive Process	Core aspect
Level II	Organizing	NOSI-6

Item SoSy12

This item is part of the item pool of the ESNaS project and therefore has to remain undisclosed.

Complexity level	Cognitive Process	Core aspect
Level III	Integrating	NOSI-6

X-Rays

Sources: Glasser, 1995; Hermann, 1972; Schreier, 2002

Stem 1

Depending on the research question data can be interpreted in various ways. Thus it may happen that different researchers gain different results from the same data. The discovery of X-rays by Wilhelm Conrad Roentgen gives an example for this.

In Roentgen's time, many physicists were interested in electron rays that are produced in glass tubes nearly free from air (see picture 1). Philipp Lenard investigated the properties of these electron rays. In Lenard's experiments, the non-visible X-rays were produced. However, he did not realize it. He just recognized that any rays made fluorescent paper glow. This sort of paper also glows if visible light meets it. Roentgen learned about Lenard's experiment and repeated it, but he covered the tube with black paper. Next to the tube, there was also some fluorescent paper that glowed during this experiment even though the electron ray tube was covered with black paper and no visible light could have escaped.

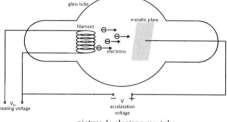

picture 1: electron ray tube

Item XR5

What was the first scientific step in Roentgen's discovery?

☑   He repeated Lenard's experiment.

☐   He tested Lenard's hypothesis.

☐   He explained Lenard's observation.

☐   He refuted Lenard's theory.

Complexity level	*Cognitive Process*	*Core aspect*
Level I	Reproducing	NOSI-5

Item XR1

After investigating the amazing properties of these rays he finally published his discovery: the existence of what he called X-rays.

What can the discovery of X-rays be cited as an example for?

☐   Scientific observational data are collected objectively so that there is only one correct result. If different results are concluded from the same data, only one of them is correct.

☑   The same observational data can lead to different conclusions. If two researchers focus on different questions they can infer two distinct scientific results from the same data.

☐   The experiments and results of scientists have to be reproducible. If two researchers conduct the same experiment at different times and places, they may not get different results.

☐   The research question determines the methods to be applied in experiments and, thus, its result. If two researchers work on the same question, they will obtain the same results.

Complexity level	*Cognitive Process*	*Core aspect*
Level V	Reproducing	NOSI-6

Stem 2

In Wilhelm Conrad Roentgen's time, many physicists were interested in electron rays that are produced in glass tubes nearly free from air (see picture 1). Philipp Lenard investigated the properties of these electron rays. In Lenard's experiments, the non-visible X-rays were produced. However, he did not realize it. He just recognized that any rays made fluorescent paper glow. This sort of paper also glows if visible light meets it. Roentgen learned about Lenard's experiment and repeated it, but he covered the tube with black paper. Next to the tube, there was also some fluorescent paper that glowed during this experiment even though the electron ray tube was covered with black paper and no visible light could have escaped.

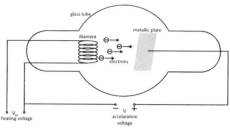

picture 1: electron ray tube

Item XR2

After investigating the amazing properties of these rays he finally published his discovery: the existence of what he called X-rays.

What can the discovery of X-rays be cited as an example for?

☐   Scientific observational data are collected objectively so that there is only one correct result. If different results are concluded from the same data, only one of them is correct.

☐   The same observational data can lead to different conclusions. If two researchers focus on different questions they can infer two distinct scientific results from the same data.

☑   The experiments and results of scientists have to be reproducible. If two researchers conduct the same experiment at different times and places, they may not get different results.

☐   The research question determines the methods to be applied in experiments and, thus, its result. If two researchers work on the same question, they will obtain the same results.

Complexity level	*Cognitive Process*	*Core aspect*
Level V	Integrating	NOSI-6

Stem 3

In Wilhelm Conrad Roentgen's time, many physicists were interested in electron rays that are produced in glass tubes nearly free from air (see picture 1). Philipp Lenard investigated the properties of these electron rays. In Lenard's experiments, the non-visible X-rays were produced. However, he did not realize it. He just recognized that any rays made fluorescent paper

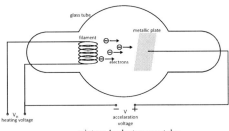

picture 1: electron ray tube

glow. This sort of paper also glows if visible light meets it. Roentgen learned about Lenard's experiment and repeated it, but he covered the tube with black paper. Next to the tube, there was also some fluorescent paper that glowed during this experiment even though the electron ray tube was covered with black paper and no visible light could have escaped.

Item XR4

First, Roentgen ensured that the covering of the tube had no leak and concluded that an unknown sort of rays must have been generated. After investigating the amazing properties of these rays he finally published his discovery: the existence of what he called X-rays.

Was Roentgen's approach scientific?

☐   No. For making the discovery, he exploited an experiment somebody else had done before.

☐   Yes. Following the scientific method he subsequently discovered the new sort of rays.

☑   Yes. Before publishing his discovery he ensured that there was no mistake in his methods.

☐   No. Striving to discover the new sort of rays, he did not change variables systematically.

Complexity level	Cognitive Process	Core aspect
Level IV	Selecting	NOSI-5

Item XR3

What can the discovery of X-rays be cited as an example for?

☐   Scientific discoveries lead to certain facts because they are objective.

☑   Scientific experiments are dependent on the research question.

☐   Scientific data cannot be interpreted differently by two researchers.

☐   Scientific theories are subject to change because they are only hypotheses.

Complexity level	Cognitive Process	Core aspect
Level III	Integrating	NOSI-4

Stem 4

In Wilhelm Conrad Roentgen's time, many physicists were interested in electron rays that are produced in glass tubes nearly free from air (see picture 1). Philipp Lenard investigated the properties of these electron rays. In Lenard's experiments, the non-visible X-rays were produced. However, he did not realize it. He just recognized that any rays made fluorescent paper

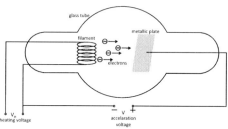

picture 1: electron ray tube

glow. This sort of paper also glows if visible light meets it. Roentgen learned about Lenard's experiment and repeated it. In doing so, he used a newly invented tube and covered it with black paper. Next to the tube, there was also some fluorescent paper that glowed during this experiment even though the electron ray tube was covered with black paper and no visible light could have escaped. He had the idea that an unknown sort of rays must have been generated. After investigating the amazing properties of these rays he finally published his discovery: the existence of what he called X-rays.

Item XR6

To what extent did Roentgen's personality influence the discovery of the X-rays? Compare it with today's research. Complete the sentence.

Scientific knowledge…

☐ … does not change due to the researcher's individual imagination. Always, the same data will be found again and again.

☐ … is true by itself. Therefore, factors like the researcher's imagination cannot affect it.

☐ … is inherent in nature. There is no need for individual imagination because the knowledge just has to be filtered out.

☑ … is not entirely objective. For example, which relationships are found depends on the researcher's individual imagination.

Complexity level	Cognitive Process	Core aspect
Level V	Integrating	NOS-1

Item XR7

This item is part of the item pool of the ESNaS project and therefore has to remain undisclosed.

Complexity level	Cognitive Process	Core aspect
Level IV	Organizing	NOS-3

Item XR8

What was the consequence of Roentgen's discovery? Complete the sentence.

The knowledge about radiation…

☐ … is true.

☑ … changed.

☐ … was created.

☐ … is certain.

Complexity level	Cognitive Process	Core aspect
Level I	Selecting	NOS-3

B.4 Location of Items in the Competence Model

Aspect "Scientific Knowledge is subjective"

Integrating			New1	---	en7; XR6
Organizing		neu8	SL7	OL3	---
Selecting	OL7	ind7; OL2	ind3	SL10	
Reproducing	ind1; OL1	Oer4	BL6; neu1	en5	BL9
NOS-1	Level I	Level II	Level III	Level IV	Level V

Aspect "Scientific Knowledge is empirically based and inferential"

Integrating			ind6	gal9	neu2
Organizing		---	BL3	SoSy3	AM2; SL3
Selecting	OL9; SoSy2	gal8	OL5	neu5	
Reproducing	neu6	en6	SoSy6	New2	OL8
NOS-2	Level I	Level II	Level III	Level IV	Level V

Aspect "Scientific Knowledge is tentative"

Integrating			OL4	SL6	neu4; nufi6
Organizing		SoSy7	Oer1	SoSy1; XR7	---
Selecting	SL13; XR8	gal1	gal2	gal5	
Reproducing	Oer2; SL4	en2	AM1	en1	AM5
NOS-3	Level I	Level II	Level III	Level IV	Level V

Aspect "Scientific investigations begin with a question"

Integrating			OI6; XR3	nufi2	SL12
Organizing		AM4	---	en8	ind2
Selecting	neu9; SL5	SL9	Oer3	SL1	
Reproducing	BL7	en9	ind5	ind8	ind4
NOSI-1	Level I	Level II	Level III	Level IV	Level V

Aspect "Scientific investigations embrace multiple methods and approaches"

Integrating			SoSy5	ind10	en3; gal11
Organizing		nufi4; SL2	BL4	---	---
Selecting	en4; SL8	gal4	AM6	XR4	
Reproducing	ind9; XR5	SoSy4	SL11	AM3	gal7
NOSI-2	Level I	Level II	Level III	Level IV	Level V

Aspect "Scientific investigations allow multiple interpretations"

Integrating			SoSy12	nufi5	BL2; XR2
Organizing		neu3; SoSy11	neu7	SoSy9	gal10
Selecting	BL1; nufi1	gal3	---	---	
Reproducing	BL5	gal6	New3	nufi3	XR1
NOSI-3	Level I	Level II	Level III	Level IV	Level V

B.5 Item Distribution on Test Booklets (Study 1)

Table B 4. Complementary Item Sets used in Study 1.

Item Set	Items
A	OL1; OL3, OL8; nufi4; nufi5; nufi6
B	gal9; gal10; gal11; BL2
C	gal1; gal3; BL6; BL8
D	Oer2; gal5, gal6, BL7; BL9
E	Oer1; gal7; gal8; BL3; BL5
F	Oer4; gal2; gal4; BL1; BL4
G	neu6; neu8; ind8; ind9; ind10
H	neu1; neu2; neu5; ind1; ind2
I	SoSy3; SoSy7; SoSy11; ind3; ind4
J	neu7; neu9; ind5; ind6; ind7
K	en1; en4; SL1; SL2; SL3; SL4
L	XR2, en2; SL8; SL9; SL10
M	XR6; en5; SL11; SL12; SL13
N	XR1; XR5; SL5; SL6; SL7
O	Oer3; AM1; AM2; AM3
P	OL4, OL9, en6, en7, en8
Q	OL2; OL6; OL7; en3
R	XR7; XR8; nufi1; nufi2; nufi3
S	OL5; AM4; AM5; AM6
T	SoSy1, SoSy2, SoSy4; SoSy12; en9
U	XR3; XR4; New1; New2; New3
V	neu3; neu4; SoSy5, SoSy6; SoSy9

Table B 5. Compilation of Test Booklets (Study 1). Each Set Appears in at Least Three Booklets.

Booklet	Item Set 1	Item Set 2	Item Set 3	# Items
1	N	P	O	14
2	V	F	M	15
3	S	B	L	13
4	I	A	C	15
5	G	O	U	14
6	Q	F	J	14
7	K	D	R	16
8	T	E	H	15
9	P	O	V	14
10	F	M	S	14
11	B	L	I	14
12	A	C	G	15
13	O	U	Q	13
14	F	J	K	16
15	D	R	T	15
16	E	H	N	15
17	O	V	M	14
18	M	S	B	13
19	L	I	A	16
20	C	G	O	13
21	U	Q	F	14
22	J	D	K	16
23	R	T	E	15
24	H	N	P	15

B.6 Item Distribution on Test Booklets (Study 2)

Table B 6. Complementary Item Sets used in Study 2.

Item Set	Items
a	Oer2; XR7; XR8; XR6; en6; en7; en8; SL5; SL6; SL7
b	New1; New2; New3; neu3; neu4; AM1; AM2; AM3
c	OL5; XR5; XR1; Oer4; en5; SL11; SL12; SL13
d	gal2; gal4; neu7; neu9; BL1; BL4; AM6; AM4; AM5
e	OL1; OL8; OL3; en3; ind3; ind4; SL8; SL9; SL10
f	nufi4; nufi6; nufi5; BL2; XR2; gal9; gal10; gal11
g	SoSy1; SoSy2; SoSy4; SoSy12; Oer3; ind8; ind9; ind10; en1; en4
h	neu1; neu2; neu5; BL6; gal1; gal3; XR4; XR3
i	nufi1; nufi2; nufi3; en2; SoSy3; SoSy7; SoSy11; ind5; ind6; ind7
j	OL6; OL7; OL2; Oer1; gal7; gal8; BL7; BL9
k	en9; ind1; ind2; SL1; SL2; SL3; SL4; neu6; neu8
l	gal5; gal6; OL4; OL9; BL3; BL5; SoSy5; SoSy6; SoSy9

Table B 7. Compilation of German Test Booklets (Study 2). Each Set Appears in Two Booklets.

Booklet	Item Set 1	Item Set 2	# Items
1	a	b	18
2	c	d	17
3	e	f	17
4	g	h	18
5	i	j	18
6	k	l	18
7	f	g	18
8	h	i	18
9	j	k	17
10	l	a	19
11	b	c	16
12	d	e	18

Table B 8. Compilation of U.S. Test Booklets (Study 2). Each Set Appears in One Booklet.

Booklet	Item Set 1	# Items
1	a	10
2	c	8
3	e	9
4	g	10
5	i	10
6	k	9
7	f	8
8	h	8
9	j	8
10	l	9
11	b	8
12	d	9

C   Statistical Data (Study 1)

C.1   Statistical Data concerning NOSSI Items

Table C 1. Detailed Item Properties Gained with the Two-dimensional Rasch Analysis (Study 1).

Item	Item Parameter	Error	Infit	Infit-T	Discrimination
AM1	1.015	0.107	1.09	2.0	.39
AM2	0.204	0.110	0.89	-2.0	.56
AM3	-0.357	0.117	1.01	0.2	.49
AM4	0.479	0.133	1.04	0.6	.45
AM5	-0.073	0.136	1.01	0.1	.47
AM6	-0.712	0.145	0.95	-0.3	.50
BL1	-0.939	0.133	0.94	-0.5	.47
BL2	0.048	0.136	1.15	1.7	.25
BL3	0.772	0.132	1.07	1.0	.40
BL4	1.173	0.114	1.02	0.5	.42
BL5		Item excluded due to misfit			
BL6		Item excluded due to misfit			
BL7	-1.420	0.158	0.90	-0.4	.53
BL9	1.000	0.132	1.07	1.1	.33
en1		Item excluded due to misfit			
en2	-1.076	0.150	1.00	0.1	.33
en3	-0.548	0.143	1.03	0.3	.33
en4	-1.454	0.158	0.83	-0.8	.40
en5	-0.871	0.137	1.01	0.1	.33
en6	-1.343	0.152	0.99	0.0	.26
en7	0.514	0.132	0.99	-0.2	.52
en8	1.928	0.138	0.98	-0.1	.28
en9	1.306	0.132	1.07	1.1	.39
gal1	-0.994	0.149	0.93	-0.4	.52
gal2	-0.540	0.125	1.00	0.1	.39
gal3	0.001	0.136	1.02	0.2	.32
gal4	-0.339	0.122	1.10	1.2	.36
gal5	0.434	0.133	1.01	0.1	.48
gal6	-0.854	0.149	1.02	0.2	.28
gal7	-0.325	0.141	1.02	0.2	.54
gal8	0.070	0.136	0.91	-1.1	.57
gal9	-0.038	0.136	0.88	-1.3	.57
gal10	0.087	0.136	0.99	-0.1	.34
gal11	1.763	0.135	1.04	0.5	.33
ind1	-1.348	0.155	1.02	0.2	.40
ind2	-0.764	0.147	0.91	-0.6	.54
ind3	-0.124	0.138	0.96	-0.4	.28
ind4	0.526	0.133	0.90	-1.7	.52

Item	Item Parameter	Error	Infit	Infit-T	Discrimination
ind5	-1.139	0.153	0.92	-0.4	.50
ind6	0.562	0.132	1.12	1.6	.28
ind7	-1.313	0.154	0.89	-0.5	.37
ind8	-0.920	0.149	0.97	-0.1	.49
ind9	-2.641	0.174	0.96	0.0	.26
ind10	1.043	0.131	1.00	0.0	.41
neu1	0.665	0.133	0.94	-0.9	.52
neu2	0.261	0.134	1.01	0.1	.48
neu3	-0.605	0.142	1.02	0.2	.47
neu4	-0.622	0.142	0.91	-0.8	.53
neu5	1.126	0.132	1.02	0.3	.35
neu6	-1.027	0.149	0.99	0.0	.49
neu7	0.783	0.132	1.00	0.0	.43
neu8	Item excluded due to misfit				
neu9	-0.145	0.138	1.12	1.2	.39
New1	1.396	0.132	1.10	1.4	.35
New2	-0.011	0.135	1.15	1.7	.21
New3	-0.572	0.143	1.00	0.0	.38
nufi1	0.025	0.136	1.02	0.2	.39
nufi2	0.398	0.133	1.03	0.5	.44
nufi3	-0.220	0.139	1.00	0.1	.34
nufi4	-0.130	0.138	0.92	-0.8	.50
nufi5	0.613	0.132	1.04	0.6	.43
nufi6	2.139	0.139	1.09	0.9	.25
Oer1	-0.702	0.145	0.92	-0.5	.51
Oer2	0.930	0.132	0.93	-1.1	.45
Oer3	-0.379	0.117	0.95	-0.6	.44
Oer4	-0.657	0.127	1.05	0.5	.34
OL1	-2.681	0.172	1.01	0.2	.26
OL2	-0.963	0.148	1.06	0.4	.30
OL3	Item excluded due to misfit				
OL4	0.298	0.133	1.06	0.9	.38
OL5	1.697	0.135	1.00	0.0	.43
OL6	0.660	0.131	1.03	0.5	.30
OL7	0.064	0.135	0.98	-0.2	.50
OL8	0.600	0.132	0.98	-0.2	.41
OL9	-0.853	0.145	0.92	-0.6	.52
SL1	-1.292	0.155	0.84	-0.8	.47
SL2	0.134	0.136	0.91	-1.1	.47
SL3	-0.099	0.137	1.00	0.0	.37
SL4	1.042	0.132	1.04	0.6	.35
SL5	-0.765	0.146	0.98	-0.1	.38
SL6	Item excluded due to misfit				
SL7	0.003	0.136	1.02	0.2	.52

Item	Item Parameter	Error	Infit	Infit-T	Discrimination
SL8	-0.725	0.146	0.97	-0.2	.47
SL9	0.602	0.133	1.00	0.0	.48
SL10	-0.328	0.140	0.98	-0.2	.55
SL11	-0.520	0.133	1.03	0.3	.38
SL12	Item excluded due to misfit				
SL13	0.480	0.122	1.02	0.3	.43
SoSy1	-0.070	0.137	0.93	-0.7	.46
SoSy2	Item excluded due to misfit				
SoSy3	0.382	0.133	0.88	-1.7	.50
SoSy4	-1.027	0.151	1.02	0.2	.34
SoSy5	0.637	0.132	1.11	1.6	.37
SoSy6	-0.670	0.142	0.96	-0.3	.40
SoSy7	0.233	0.134	0.94	-0.8	.58
SoSy9	0.637	0.132	0.99	-0.2	.46
SoSy11	0.161	0.135	1.10	1.3	.25
SoSy12	2.138	0.139	1.09	0.9	.27
XR1	0.668	0.132	1.02	0.3	.45
XR2	0.707	0.132	1.02	0.4	.43
XR3	Item excluded due to misfit				
XR4	1.709	0.135	1.00	0.0	.33
XR5	0.564*	0.960	0.99	-0.2	.40
XR6	0.871	0.122	1.01	0.1	.46
XR7	-1.381	0.155	0.96	-0.1	.34
XR8	1.027*	0.942	1.02	0.4	.39

* Item parameter was constrained.

Table C 2. Levene-Test for Factors Used in ANOVA (Study 1).

Factor	F	df1	df2	p	Homogeneity of Variances
Complexity Level	1.47	4	92	.22	✓
Cognitive Process	1.74	3	93	.16	✓
Core Aspect	0.89	5	91	.49	✓

Table C 3. Kolmogorov-Smirnov-Test for Each Group of Complexity Levels Used for ANOVA (Study 1).

Group	n	μ	s.d.	Z	p*	Normal Distribution
Level I	18	-0.55	1.14	0.50	.97	✓
Level II	19	-0.33	0.73	0.57	.90	✓
Level III	22	0.29	0.89	0.65	.79	✓
Level IV	19	0.15	0.92	0.41	1.00	✓
Level V	19	0.37	0.75	0.50	.96	✓

* two-tailed

Table C 4. Kolmogorov-Smirnov-Test for Each Group of Cognitive Processes Used for ANOVA (Study 1).

Group	n	μ	s.d.	Z	p*	Normal Distribution
reproducing	31	-0.37	1.03	0.73	.67	✓
selecting	27	-0.17	0.87	0.68	.75	✓
organizing	20	0.16	0.73	0.65	0.80	✓
integrating	19	0.68	0.77	0.73	.66	✓

* two-tailed

Table C 5. Kolmogorov-Smirnov-Test for Each Group of Core Aspects Used for ANOVA (Study 1).

Group	n	μ	s.d.	Z	p*	Normal Distribution
NOS-1	15	-0.25	1.08	0.36	1.00	✓
NOS-2	16	0.10	0.80	0.60	.86	✓
NOS-3	17	0.13	0.94	0.47	.98	✓
NOSI-4	15	-0.06	1.00	0.62	.84	✓
NOSI-5	18	-0.10	1.12	0.58	.89	✓
NOSI-6	16	0.17	0.77	0.60	.87	✓

* two-tailed

C.2 Person characteristics (Study 1)

Table C 6. Comparison of median values P_{50} and mean values μ with standard deviation σ of the observed person characteristics.

Instrument	Scale/Variable	P_{50}	μ	σ
NOSSI	Person parameter (NOSI)	0.810	0.886	1.167
	Person parameter (NOS)	0.907	0.842	1.167
KFT-N2	Non-verbal cognitive ability	37.000	38.253	8.843
LGVT	Reading comprehension	51.000	51.210	11.043
	Reading speed	52.000	53.157	10.566
SIS$_{ad}$	FI	2.143	2.179	0.761
	SI	1.667	1.811	0.606
	AI	1.600	1.732	0.742
	EK	2.500	2.448	0.750
	SK	2.429	2.421	0.766
	SW	2.400	2.325	0.785
SNOS	Q	4.167	4.047	0.717
	S	3.750	3.706	0.569
	E	4.000	3.925	0.565
	R	4.091	4.018	0.419
	K	3.000	3.066	0.628
	Z	3.857	3.794	0.470
	V	3.286	3.246	0.631

C.3 Statistical Data concerning KFT-N2

T-values regarding non-verbal cognitive abilities were not normal distributed, $Z(1067) = 2.90, \ p < .01$.

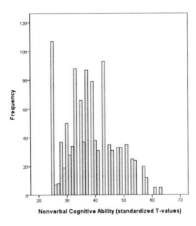

Figure C 1. Distribution of observed cognitive abilities.

C.4 Statistical Data concerning LGVT

T-values regarding reading speed were not normal distributed, $Z(996) = 2.88$, $p < .01$.

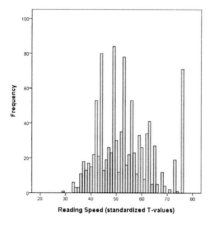

Figure C 2. Distribution of observed reading speed.

T-values regarding reading comprehension were not normal distributed, $Z(1063) = 2.06$, $p < .01$.

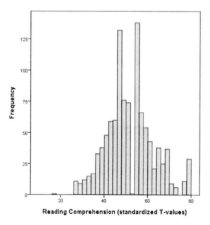

Figure C 3. Distribution of observed reading comprehension.

C.5 Statistical Data concerning SIS_{ad}

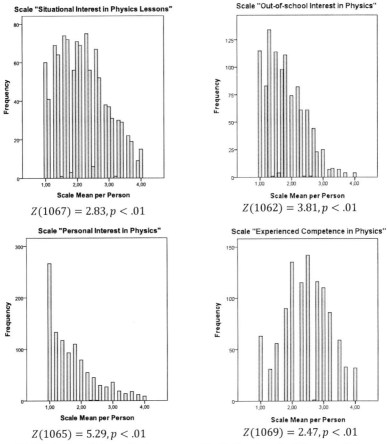

Figure C 4. Distributions of observed scale means as well as findings from the Kolmogorov-Smirnov test for SIS_{ad} scales 'Situational Interest', 'Out-of-school Interest', 'Personal Interest' and 'Experienced Competence'.

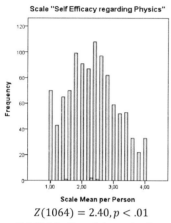 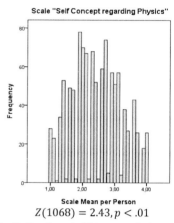

$$Z(1064) = 2.40, p < .01 \qquad\qquad Z(1068) = 2.43, p < .01$$

Figure C 5. Distributions of observed scale means as well as findings from the Kolmogorov-Smirnov test for SIS_{ad} scales 'Self efficacy' and 'Self Concept'.

C.6 Statistical Data concerning SNOS

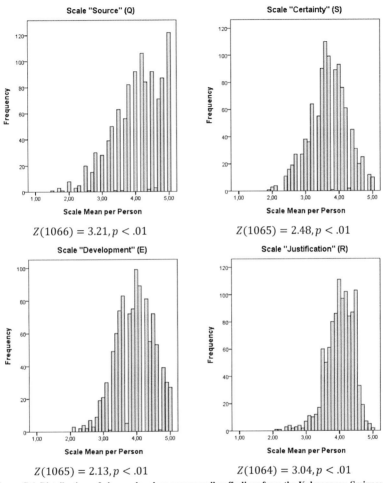

$Z(1066) = 3.21, p < .01$

$Z(1065) = 2.48, p < .01$

$Z(1065) = 2.13, p < .01$

$Z(1064) = 3.04, p < .01$

Figure C 6. Distributions of observed scale means as well as findings from the Kolmogorov-Smirnov test for SNOS scales 'Source', 'Certainty', 'Development', and 'Justification'.

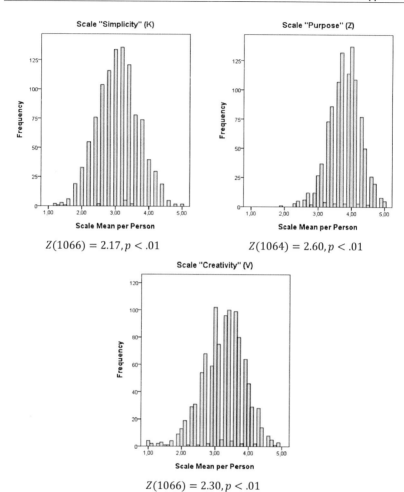

$$Z(1066) = 2.17, p < .01 \qquad Z(1064) = 2.60, p < .01$$

$$Z(1066) = 2.30, p < .01$$

Figure C 7. Distributions of observed scale means as well as findings from the Kolmogorov-Smirnov test for SNOS scales 'Simplicity', 'Purpose', and 'Creativity'.

C.7 Comparison of Correlations

Table C 7. Confidence Intervals of Pearson's product-moment correlation between NOSI competence and competence-related constructs.

Correlation with	Confidence intervals	
	Lower bound	Upper bound
nonverbal cognitive ability (KFT-N2)	.153	.342
reading comprehension (LV)	.174	.361
reading speed (LG)	-.004	.202
situational interest (FI)	not significant	
out-of-school interest (AI)	not significant	
personal interest (SI)	-.041	.160
experienced competence (EK)	.000	.198
self concept (SK)	.071	.266
self efficacy (SW)	.081	.275
Source (Q)	.122	.314
Certainty (S)	.132	.323
Development (E)	.185	.370
Justification (R)	.132	.323
Simplicity (K)	.071	.266
Purpose (Z)	.009	.208
Creativity (V)	-.031	.169

Table C 8. Confidence Intervals of Pearson's product-moment correlation between NOS competence and competence-related constructs.

Correlation with	Confidence intervals	
	Lower bound	Upper bound
nonverbal cognitive ability (KFT-N2)	.174	.361
reading comprehension (LV)	.195	.380
reading speed (LG)	.047	.250
situational interest (FI)	not significant	
out-of-school interes (AI)	not significant	
personal interest (SI)	-.011	.189
experienced competence (EK)	-.021	.179
self concept (SK)	.071	.266
self efficacy (SW)	.091	.285
Source (Q)	.143	.333
Certainty (S)	.153	.342
Development (E)	.248	.426
Justification (R)	.153	.342
Simplicity (K)	.132	.323
Purpose (Z)	-.011	.189
Creativity (V)	-.001	.199

Table C 9. Comparison of Correlations of Competence Concerning NOSI and SNOS Scales with Correlations of Competence Concerning NOSI and Remaining Constructs.

	SNOS Scales						
	Q	S	E	R	K	Z	V
non-verbal cognitive ability	n.s.	n.s.	n.s.	n.s.	n.s.	n.s.	n.s.
reading comprehension	n.s.	n.s.	n.s.	n.s.	n.s.	n.s.	-
reading speed	n.s.	n.s.	n.s.	n.s.	n.s.	n.s.	n.s.
situational interest				not compared			
out-of-school interes				not compared			
personal interest	n.s.	n.s.	+	n.s.	n.s.	n.s.	n.s.
experienced competence	n.s.	n.s.	n.s.	n.s.	n.s.	n.s.	n.s.
self concept	n.s.	n.s.	n.s.	n.s.	n.s.	n.s.	n.s.
self efficacy	n.s.	n.s.	n.s.	n.s.	n.s.	n.s.	n.s.

Table C 10. Comparison of Correlations of Competence Concerning NOS and SNOS Scales with Correlations of Competence Concerning NOS and Remaining Constructs.

	SNOS Scales						
	Q	S	E	R	K	Z	V
non-verbal cognitive ability	n.s.	n.s.	n.s.	n.s.	n.s.	n.s.	n.s.
reading comprehension	n.s.	n.s.	n.s.	n.s.	n.s.	-	n.s.
reading speed	n.s.	n.s.	n.s.	n.s.	n.s.	n.s.	n.s.
situational interest				not compared			
out-of-school interest				not compared			
personal interest	n.s.	n.s.	+	n.s.	n.s.	n.s.	n.s.
experienced competence	n.s.	n.s.	+	n.s.	n.s.	n.s.	n.s.
self concept	n.s.	n.s.	n.s.	n.s.	n.s.	n.s.	n.s.
self efficacy	n.s.	n.s.	n.s.	n.s.	n.s.	n.s.	n.s.

D   Statistical Data (Study 2)

D.1 Item characteristics (Study 2)

Table D 1. Detailed Item Properties Gained with the Two-dimensional Rasch Analysis (Study 2).

Item	Item Parameter	Error	Infit	Infit-T	Discrimination
AM1	Item excluded due to misfit				
AM2	0.494	0.116	1.09	1.2	.39
AM3	0.140	0.130	0.92	-1.2	.57
AM4	0.247	0.128	0.99	-0.1	.48
AM5	-0.891	0.121	1.05	0.5	.40
AM6	-0.661	0.135	0.89	-1.2	.56
BL1	-0.697	0.135	0.96	-0.4	.51
BL2	0.132	0.128	1.07	1.2	.33
BL3	0.136	0.116	1.02	0.3	.48
BL4	1.629	0.134	1.09	1.0	.23
BL5	1.440	0.132	1.03	0.4	.43
BL6	-0.782	0.120	1.06	0.6	.34
BL7	-1.144	0.142	0.91	-0.7	.56
BL9	0.117	0.116	1.01	0.1	.52
en1	-0.068	0.116	0.96	-0.5	.55
en2	-0.855	0.120	0.97	-0.3	.57
en3	0.206	0.127	1.07	1.2	.38
en4	-0.738	0.135	1.05	0.5	.43
en5	-0.355	0.116	0.92	-1.0	.55
en6	-0.697	0.108	1.01	0.1	.54
en7	1.129	0.105	0.95	-0.9	.56
en8	Item excluded due to misfit				
en9	Item excluded due to misfit				
gal1	-1.053	0.122	0.91	-0.9	.58
gal2	-0.969	0.122	0.99	-0.1	.49
gal3	-0.660	0.135	0.99	-0.1	.47
gal4	-0.522	0.133	1.06	0.8	.40
gal5	0.043	0.116	0.96	-0.5	.57
gal6	-0.709	0.136	1.01	0.2	.39
gal7	-0.487	0.134	0.97	-0.3	.54
gal8	0.725	0.115	1.02	0.3	.44
gal9	0.337	0.114	0.95	-0.8	.55
gal10	0.349	0.127	0.98	-0.3	.50
gal11	Item excluded due to misfit				
ind1	-0.634	0.119	1.01	0.2	.46
ind2	-0.310	0.132	0.91	-1.2	.55
ind3	-0.130	0.115	0.96	-0.6	.50
ind4	-0.055	0.128	0.96	-0.7	.50
ind5	0.072	0.128	1.05	0.8	.51
ind6	Item excluded due to misfit				

Item	Item Parameter	Error	Infit	Infit-T	Discrimination
ind7	Item excluded due to misfit				
ind8	-0.632	0.134	1.02	0.3	.54
ind9	-2.190	0.159	0.92	-0.4	.36
ind10	0.186	0.128	1.05	0.8	.43
neu1	0.722	0.115	1.00	0.0	.39
neu2	Item excluded due to misfit				
neu3	-0.048	0.131	0.93	-1.0	.53
neu4	-0.482	0.119	0.94	-0.7	.59
neu5	Item excluded due to misfit				
neu6	-1.186	0.124	1.03	0.3	.50
neu7	0.908	0.128	1.10	1.6	.39
neu8	Item excluded due to misfit				
neu9	-0.455	0.132	1.02	0.2	.42
New1	1.325	0.118	1.09	1.0	.36
New2	-0.010	0.117	1.02	0.2	.54
New3	-0.483	0.135	0.95	-0.6	.50
nufi1	-0.017	0.129	1.03	0.4	.55
nufi2	0.391	0.127	1.03	0.5	.47
nufi3	0.042	0.128	1.01	0.2	.46
nufi4	0.161	0.128	0.91	-1.4	.55
nufi5	0.132	0.128	1.08	1.3	.33
nufi6	1.337	0.117	1.07	1.0	.31
Oer1	-0.396	0.118	0.89	-1.4	.56
Oer2	Item excluded due to misfit				
Oer3	-0.175	0.130	0.95	-0.7	.53
Oer4	-0.517	0.117	1.02	0.3	.48
OL1	Item excluded due to misfit				
OL2	-1.244	0.125	0.87	-1.0	.56
OL3	Item excluded due to misfit				
OL4	-0.114	0.116	0.86	-2.0	.57
OL5	-0.652	0.118	0.96	-0.5	.50
OL6	Item excluded due to misfit				
OL7	-0.752	0.120	1.06	0.7	.52
OL8	-0.410	0.117	1.05	0.6	.45
OL9	-0.583	0.119	0.83	-2.0	.59
SL1	-1.265	0.144	0.83	-1.3	.64
SL2	-0.277	0.132	0.96	-0.5	.56
SL3	-0.157	0.117	0.95	-0.6	.57
SL4	1.613	0.120	1.09	1.0	.39
SL5	0.343	0.112	0.91	-1.9	.64
SL6	Item excluded due to misfit				
SL7	0.670	0.103	1.00	0.1	.51
SL8	0.062	0.128	1.08	1.3	.36
SL9	0.856	0.127	0.95	-0.9	.47
SL10	0.167	0.114	0.94	-1.0	.46

Item	Item Parameter	Error	Infit	Infit-T	Discrimination
SL11	0.434	0.127	0.96	-0.6	.47
SL12	Item excluded due to misfit				
SL13	1.188	0.117	1.07	0.9	.35
SoSy1	Item excluded due to misfit				
Sosy2	-0.006	0.115	0.96	-0.6	.54
SoSy3	Item excluded due to misfit				
SoSy4	Item excluded due to misfit				
SoSy5	Item excluded due to misfit				
SoSy6	-0.376	0.118	1.01	0.2	.49
SoSy7	1.048	0.115	1.05	0.8	.43
SoSy9	0.440	0.128	1.06	1.0	.41
SoSy11	-0.138	0.130	0.92	-1.0	.53
SoSy12	Item excluded due to misfit				
XR1	0.463	0.127	1.00	-0.1	.47
XR2	0.591	0.127	1.02	0.4	.41
XR3	1.770	0.137	1.20	2.0	.26
XR4	1.519	0.134	1.11	1.3	.31
XR5	-0.846*	0.863	0.94	-0.7	.52
XR6	1.067	0.105	0.90	-1.7	.56
XR7	-0.434	0.106	0.95	-0.8	.51
XR8	1.634*	0.745	1.05	0.7	.39

* Item parameter was constrained.

Table D 2. Levene-Test for Factors Used in ANOVA (Study 2).

Factor	F	df1	df2	p	Homogeneity of Variances
Complexity Level	2.60	4	81	.04	X
Cognitive Process	0.67	3	82	.57	✓
Core Aspect	1.68	5	80	.15	✓

Table D 3. Kolmogorov-Smirnov-Test for Each Group of Complexity Levels Used for ANOVA (Study 2).

Group	n	μ	s.d.	Z	p*	Normal Distribution
Level I	18	-0.16	1.06	0.69	.72	✓
Level II	16	-0.23	0.69	0.65	.79	✓
Level III	19	0.15	0.82	0.67	.76	✓
Level IV	16	0.04	0.59	0.74	.64	✓
Level V	17	0.18	0.62	0.41	1.00	✓

* two-tailed

Table D 4. Kolmogorov-Smirnov-Test for Each Group of Cognitive Processes Used for ANOVA (Study 2).

Group	n	μ	s.d.	Z	p*	Normal Distribution
reproducing	29	-0.29	0.77	0.66	.78	✓
selecting	26	-0.16	0.80	0.74	.64	✓
organizing	17	0.25	0.57	0.47	.98	✓
integrating	14	0.57	0.65	0.68	.75	✓

* two-tailed

Table D 5. Kolmogorov-Smirnov-Test for Each Group of Core Aspects Used for ANOVA (Study 2).

Group	n	μ	s.d.	Z	p*	Normal Distribution
NOS-1	14	0.06	0.82	0.49	.97	✓
NOS-2	13	-0.18	0.54	0.37	1.00	✓
NOS-3	15	0.11	0.98	0.75	.64	✓
NOSI-4	13	-0.03	0.81	0.54	.94	✓
NOSI-5	15	-0.09	0.93	0.68	.75	✓
NOSI-6	16	0.11	0.59	0.44	.99	✓

* two-tailed

D.2 Person characteristics (Study 2)

Table D 6. Comparison of median values P_{50} and mean values μ with standard deviation σ of the observed NOSI and NOS person parameters for both subsamples.

Subsample	Scale/Variable	P_{50}	μ	σ
German students	Person parameter (NOSI)	0.570	0.695	1.163
	Person parameter (NOS)	0.618	0.669	1.249
U.S. students	Person parameter (NOSI)	-0.282	-0.136	1.232
	Person parameter (NOS)	-0.138	-0.262	1.249

List of Figures

List of Tables

References

Abd-El-Khalick, F., & Lederman, N. G. (2000). The influence of history of science courses on students' views of nature of science. *Journal of Research in Science Teaching 37*(10), 1057-1095.

Achilles, M. (1996). *Historische Versuche der Physik, nachgebaut und kommentiert.* Frankfurt am Main: Wötzel.

Adams, R. (2002). Scaling PISA cognitive data. In R. Adams & M. Wu (Eds.), *PISA 2000 technical report* (99-108). Paris: OECD.

Adams, R. J., & Khoo, S. H. (1996). *Quest.* Melbourne: Australian Council for Educational Research.

Alters, B. J. (1997). Whose nature of science? *Journal of Research in Science Teaching 34*(1), 39-55.

American Association for the Advancement of Science [AAAS] (1990). Science for all Americans. New York, NY: Oxford University Press.

American Association for the Advancement of Science [AAAS] (1993). Benchmarks for science literacy. New York, NY: Oxford University Press.

American Educational Research Association [AERA], American Psychological Association [APA], & National Council on Measurement in Education [NCME] (2004). *Standards for educational and psychological testing.* Washington, DC, American Educational Research Association.

Arbuckle, J. L. (2009). Amos [computer software]. Crawfordville, FL: Amos Development Corporation.

Baumert, J. (2002). Deutschland im internationalen Bildungsvergleich. In N. Killius, J. Kluge & L. Reisch (Eds.), *Die Zukunft der Bidung* (100-150). Frankfurt am Main: Suhrkamp.

Beauducel, A., & Wittmann, W. W., (2005). Simulation study on fit indexes in CFA based on data with slightly distorted simple structure. *Structural Equation Modeling: A Multidisciplinary Journal 12*(1), 41 – 75.

Bernholt, S. (2010). *Kompetenzmodellierung in der Chemie: Theoretische und empirische Reflexion am Beispiel des Modells hierarchischer Komplexität.* Berlin: Logos.

Bernholt, S., Parchmann, I., & Commons, M. L. (2009). Kompetenzmodellierung zwischen Forschung und Unterrichtspraxis. *Zeitschrift für Didaktik der Naturwissenschaften 15*, 219-245.

Best, J. W., & Kahn, J. V. (2006). *Research in education.* Boston: Pearson.

Biesta, G. (2002). How general can *Bildung* be? Reflections on the future of a modernb educational ideal. *Journal of Philosophy of Education, 36*(3), 377-390.

Bond, T. G., & Fox, C. M. (2007). *Applying the Rasch model.* Mahwah, NJ: Erlbaum.

Bortz, J. (2005). *Statistik für Human- und Sozialwissenschaftler.* Heidelberg: Springer.

Bortz, J., & Döring, N. (2006). *Forschungsmethoden und Evaluation für Human- und Sozialwissenschaftler.* Heidelberg: Springer.

Bühner, M. (2006). *Einführung in die Test- und Fragebogenkonstruktion.* München: Pearson.

Bybee, R. W. (1997). Toward an understanding of scientific literacy. In W. Gräber & C. Bolte (Eds.), *Scientific Literacy: An international symposium* (37-68). Kiel: Institute of Science Education (IPN).

Bybee, R., McCrae, B., & Laurie, R. (2009). PISA 2006: An assessment of scientific literacy. *Journal of Research in Science Teaching 46*(8), 865-883.

Carey, S., Evans, R., Honda, M., Jay, E., Unger, C. (1989). 'An experiment is when you try it and see if it works': A study of grade 7 students' understanding of the construction of scientific knowledge. *International Journal of Science Education 11*(5), 514-529.

Chadwick, J. (1932a). Possible existence of a neutron. *Nature 129*(3252), 312.

Chadwick, J. (1932b). Existence of a neutron. *Proceedings of the Royal Society A (136),* 692-708.

Chalmers, A. F. (2007). *Wege der Wissenschaft: Einführung in die Wissenschaftstheorie* (6th ed.). Berlin: Springer.

Chomsky, N. (1986). *Language and mind.* New York, NY: Harcourt, Brace & World.

Cohen, J. (1988). *Statistical power analysis for the behavioral sciences.* Hillsdale, NJ: Lawrence Erlbaum Associates.

Cohen, J. (1992). A power primer. *Psychological Bulletin 112*(1), 155-159.

Commons, M. L., Trudeau, E. J., Stein, S. A., Richards, F. A., & Krause, S. R. (1998). Hierarchical complexity of tasks shows the existence of developmental stages. *Developmental Review 18*(3), 237-278.

DeBoer, G. E. (1997). Historical perspectives on scientific literacy. In W. Gräber & C. Bolte (Eds.), *Scientific Literacy: An international symposium* (69-86). Kiel: Institute of Science Education (IPN).

de Boer, K. S. (2007). *Wie konnte Hubble zeigen, dass das Universum expandiert?* Retrieved from http://www.astro.uni-bonn.de/~deboer/hubble/hubble.html

Deger, H., Gleixner, C., Pippig, R., & Worg, R. (2001). *Galileo: das anschauliche Physikbuch*. München: Oldenbourg.

Döbert, H. (2007). Germany. In W. Hörner, H. Döbert, B. von Kopp & W. Mitter (Eds.), *The education systems of Europe* (299-325). Dordrecht: Springer.

Driver, R., Leach, J., Millar, R., & Scott, P. (1996). *Young people's images of science*. Buckingham: Open University Press.

Einhaus, E. (2007). *Schülerkompetenzen im Bereich Wärmelehre*. Berlin: Logos.

Fermi, E. (1934a). Radioactivity induced by neutron bombardment. *Nature 133*(3368), 757.

Fermi, E. (1934b). Possible production of elements of atomic number higher than 92. *Nature 133*(3372), 898-899.

Feyerabend, P. (1993). *Against method* (3rd ed.). London: Verso.

Field, A. (2009). *Discovering statistics using SPSS*. London: Sage Publications.

Fischer, H. E., Glemnitz, I., Kauertz, A., & Sumfleth, E. (2007). Auf Wissen aufbauen – kumulatives Lernen in Chemie und Physik. In E. Kircher, R. Girwidz & P. Häußler (Eds.), *Physikdidaktik: Theorie und Praxis* (657-678). Berlin: Springer.

Fischer, H. E., Kauertz, A., & Neumann, K. (2008). Standards and Bildung. In S. Mikelskis-Seifert, U. Ringelband & M. Brückmann (Eds.), *Four decades of research in science education – from curriculum development to quality improvement* (29-41). Münster: Waxmann.

Fraunberger, F. (1985). *Illustrierte Geschichte der Elektrizität*. Köln: Aulis Verlag Deubner.

Furr, R. M., & Bacharach, V. R. (2008). *Psychometrics: An introduction*. Los Angeles: Sage.

Galili, I., & Hazan, A. (2001). The effect of a history-based course in optics on students' views about science. *Science & Education 10*(1-2), 7-32.

Geiger, H., & Marsden, E. (1909). On a diffuse reflection of the α-particles. *Proceedings of the Royal Society A 82*(557), 495-500.

Glasser, O. (1995). *Wilhelm Conrad Röntgen und die Geschichte der Röntgenstrahlen.* Heidelberg: Springer Verlag.

Grygier, P. (2008). *Wissenschaftsverständnis von Schülern im Sachunterricht.* Bad Heilbrunn: Klinkhardt.

Haertel, E. H. (2006). Reliability. In R. L. Brennan (Ed.), *Educational Measurement* (65-110). Westport, CT: American Council on Education [ACE] and Praeger.

Hakim, J. (2004). *The Story of Science. Aristotle leads the way.* Washington: Smithsonian Books.

Hakim, J. (2005). *The Story of Science. Newton at the Center.* Washington: Smithsonian Books.

Halloun, I. A. (2001). *Student views about science: A comparative survey.* Beirut: Educational Research Center, Lebanese University.

Halloun, I. A., & Hestenes, D. (1996). *Views about Science Survey: VASS.* Paper presented at the annual meeting of the National Association for Research in Science Teaching, Saint Louis, MO. ERIC document No. ED394840.

Halloun, I. A., & Hestenes, D. (1998). Interpreting VASS dimensions and profiles for physics students. *Science & Education 7*(6), 553-577.

Hammann, M. (2004). Kompetenzentwicklungsmodelle: Merkmale und ihre Bedeutung – dargestellt anhand von Kompetenzen beim Experimentieren. *Der mathematische und naturwissenschaftliche Unterricht 57*(4), 196-203.

Harré, R. (1984). *Great scientific experiments.* Oxford: Oxford University Press.

Hartig, J., & Klieme, E. (2006). Kompetenz und Kompetenzdiagnostik. In K. Schweizer (Ed.), *Leistung und Leistungsdiagnostik* (127-143). Berlin: Springer.

Hawking, S. W. (1988). *Eine kurze Geschichte der Zeit.* Reinbek: Rowohlt.

Heering, P. (2001). Ohms Drehwaage und der elektrische Widerstand. *Physik in unserer Zeit 32*(2), 90- 91.

Heller, A. (1965). *Geschichte der Physik von Aristoteles bis auf die neueste Zeit: Von Decartes bis Robert Mayer.* Stuttgart: Enke.

Heller, K. A., & Perleth, C. (2000). *Kognitiver Fähigkeitstest für 4.-12. Klassen, Revision (KFT 4-12+ R).* Göttingen: Hogrefe.

Helmke A. (2000). TIMSS und die Folgen: Der weite Weg von der externen Leistungsevaluation zur Verbesserung des Lehrens und Lernens. In U. P. Trier (Ed.), *Bildungswirksamkeit zwischen Forschung und Politik* (135-164). Chur: Rüegger.

Hermann, A. (1972). *Lexikon Geschichte der Physik: Biographien, Sachwörter, Original-schriften und Sekundärliteratur.* Köln: Aulis Verlag Deubner.

Hofer, B. K., & Pintrich, P. R. (1997). The development of epistemological theories: Beliefs about knowledge and knowing and their relation to learning. *Review of Educational Research 67*(1), 88-140.

Hößle, C., Höttecke, D., & Kircher, E. (Eds.). (2004). *Lehren und lernen über die Natur der Naturwissenschaften.* Baltmannsweiler: Schneider Verlag Hohengehren.

Höttecke, D. (2001a). Die Vorstellungen von Schülerinnen und Schülern von der "Natur der Naturwissenschaften". *Zeitschrift für Didaktik der Naturwissenschaften 7*, 7-23.

Höttecke, D. (2001b). *Die Natur der Naturwissenschaften historisch verstehen: Fachdidaktische und wissenschaftshistorische Untersuchungen.* Berlin: Logos.

Höttecke, D. (2004a). Wissenschaftsgeschichte im naturwissenschaftlichen Unterricht. In C. Hößle, D. Höttecke & E. Kircher (Eds.), *Lehren und lernen über die Natur der Naturwissenschaften* (43-56). Baltmannsweiler: Schneider Verlag Hohengehren.

Höttecke, D. (2004b). Schülervorstellungen über die "Natur der Naturwissenschaften". In C. Hößle, D. Höttecke & E. Kircher (Eds.), *Lehren und lernen über die Natur der Naturwissenschaften* (264-277). Baltmannsweiler: Schneider Verlag Hohengehren.

Höttecke, D., & Rieß, F. (2009). *Developing and implementing case studies for teaching science with the help of history and philosophy: Framework and critical perspectives on "HIPST" - a european approach for the inclusion of history and philosophy in science teaching.* Paper presented at the Tenth International History, Philosophy, and Science Teaching Conference, South Bend, IN.

Hund, F. (1972). *Geschichte der physikalischen Begriffe.* Mannheim: Bibliographisches Institut.

Jorda, S. (2008). Kalt und kostbar. *Physik Journal 7*(7), 27- 30.

Jude, N., Klieme, E., Eichler, W., Lehmann, R. H., Nold, G., Schröder, K., Thomé, G., & Willenberg, H. (2008). Strukturen sprachlicher Kompetenzen. In DESI-Konsortium (Ed.), *Unterricht und Kompetenzerwerb in Deutsch und Englisch* (191-201). Weinheim: Beltz.

Kane, M. T. (2006). Validation. In R. L. Brennan (Ed.), *Educational measurement* (17-64). Westport, CT: American Council on Education and Praeger.

Kant, I. (1992). An answer to the question: What is enlightenment? In P. Waugh (Ed.) *Postmodernism: A reader* (89-95). London: Arnold.

Kauertz, A. (2008). *Schwierigkeitserzeugende Merkmale physikalischer Leistungstestaufgaben.* Berlin: Logos.

Kauertz, A. (2009, September). *Effekte interessensorientierten Unterrichts in der Primarstufe: Operationalisierung von Interesse und Interessensorientierung und Pilotierung der Instrumente.* Paper presented at the meeting of the Gesellschaft für Didaktik der Chemie und Physik, Dresden.

Kauertz, A., & Fischer, H. E. (2006). Assessing students' level of knowledge and analysing the reasons for learning difficulties in physics by Rasch analysis. In X. Liu & W. Boone (Eds.), *Applications of Rasch measurement in science education* (212-246). Maple Grove, MA: Jam Press.

Kauertz, A., Fischer, H. E., Mayer, J., Sumfleth, E., & Walpuski, M. (2010). Standardbezogene Kompetenzmodellierung in den naturwissenschaftlichen Fächern der Sekundarstufe I. *Zeitschrift für Didaktik der Naturwissenschaften.*

Kinnebrock, W. (2002). *Bedeutende Theorien des 20. Jahrhunderts: Ein Vorstoß zu den Grenzen von Berechenbarkeit und Erkenntnis.* München: Oldenbourg.

Kipnis, N. (1998). A history of science approach to the nature of science: learning science by rediscovering it. In W. F. McComas (Ed.), *The nature of science in science education: Rationales and strategies* (177-196). Dordrecht: Kluwer.

Kircher, E. (2007). Warum Physikunterricht? In E. Kircher, R. Girwidz & P. Häußler (Eds.), *Physikdidaktik: Theorie und Praxis* (13-78). Berlin: Springer.

Kircher, E., & Dittmer, A. (2004). Lehren und lernen über die Natur der Naturwissenschaften – ein Überblick –. In C. Hößle, D. Höttecke & E. Kircher (Eds.), *Lehren und lernen über die Natur der Naturwissenschaften* (2-22). Baltmannsweiler: Schneider Verlag Hohengehren.

Kishfe, R., & Abd-El-Khalick, F. (2002). Influence of explicit and reflective versus implicit inquiry-oriented instruction on sixth graders' views of nature of science. *Journal of Research in Science Teaching 39*(7), 551-578.

Klafki, W. (2007). *Neue Studien zur Bildungstheorie und Didaktik: Zeitgemäße Allgemeinbildung und kritisch-konstruktive Didaktik* (6th ed.). Weinheim: Beltz.

Kleinert, A. (Ed.). (1980). *Physik im 19. Jahrhundert.* Darmstadt: Wissenschaftliche Buchgesellschaft.

Klieme, E., Avenarius, H., Blum, W., Döbrich, P., Gruber, H., Prenzel, M., Reiss, K., Riquarts, K., Rost, J., Tenorth, H.-E., & Vollmer, H. J. (2004). *The development of national educational standards: An expertise.* Berlin: Federal Ministry of Education and Research (BMBF).

Klieme, E., & Hartig, J. (2007). Kompetenzkonzepte in den Sozialwissenschaften und im erziehungswissenschaftlichen Diskurs. In M. Prenzel, I. Gogolin, & H.-H. Krüger (Eds.), *Kompetenzdiagnostik: Zeitschrift für Erziehungswissenschaft Sonderheft 8/2007* (11-29). Wiesbaden: VS Verlag für Sozialwissenschaften.

Klieme, E., Hartig, J., & Rauch, D. (2008). The concept of competence in educational contexts. In J. Hartig, E. Klieme & D. Leutner (Eds.), *Assessment of competencies in educational contexts* (3-22). Cambrige, MA: Hogrefe.

Klieme, E., & Leutner, D. (2006). Kompetenzmodelle zur Erfassung individueller Lernergebnisse und zur Bilanzierung von Bildungsprozessen: Beschreibung eines neu eingerichteten Schwerpunktprogramms der DFG. *Zeitschrift für Pädagogik 52*, 876-903.

Klopfer, L. E. (1969). Science Education in 1991. *The School Review 77*(3/4), 199-217.

Klopfer, L. E., & Cooley, W. W. (1963). The *history of science cases for high schools* in the development of student understanding of science and scientists: A report on the HOSC instruction project. *Journal of Research in Science Teaching 1*(1), 33-47.

Koeppen, K., Hartig, J., Klieme, E., & Leutner, D. (2008). Current issues in competence modeling and assessment. *Zeitschrift für Psychologie/ Journal of Psychology 216*(2), 61-73.

Köller, O., Katzenbach, M., Mayer, J., Hartmann, S., Kremer, K., Wellnitz, N. Sumfleth, E., Walpuski, M., Ropohl, M., Fischer, H. E., Kauertz, A., Notarp, H., Zilker, I. (2008). *Aufgabenkonstruktionsanleitung: Aufgabenentwicklung in den Fächern Biologie, Chemie und Physik für den Kompetenzbereich „Erkenntnisgewinnung".* Unpublished paper in scope of the project ESNaS.

Kuhn, T. S. (1976). *Die Struktur wissenschaftlicher Revolutionen.* Frankfurt am Main: Suhrkamp.

Kulgemeyer, C., & Schecker, H. (2009). Kommunikationskompetenz in der Physik: Zur Entwicklung eines domänenspezifischen Kommunikationsbegriffs. *Zeitschrift für Didaktik der Naturwissenschaften 15*, 131-153.

Lakatos, I. (1978). *The methodology of scientific research programmes.* Cambridge: Cambridge University Press.

Lederman, N. G. (2006). Syntax of nature of science within inquiry and science instruction. In L. B. Flick & N. G. Lederman (Eds.), *Scientific inquiry and nature of science: Implications for teaching, learning and teacher education* (301-317). Dordrecht: Springer.

Lederman, N. G. (2007). Nature of science: Past, present, and future. In S. K. Abell & N. G. Lederman (Eds.), *Handbook of research on science education* (831-879). Mahwah, NJ: Erlbaum.

Lederman, N. G., & Abd-El-Khalick, F. (1998). Avoiding de-natured science: Activities that promote understandings of the nature of science. In W. F. McComas (Ed.), *The nature of science in science education: Rationales and strategies* (83-126). Dordrecht: Kluwer.

Lederman, N. G., Abd-El-Khalick, F., Bell, R. L., & Schwartz, R. S. (2002). Views of Nature of Science quesionnaire: Toward valid and meaningful assessment of learner's conceptions of nature of science. *Journal of Research in Science Teaching* 39(6), 497-521.

Lederman, N. G., Wade, P., & Bell, R. L. (1998). Assessing understanding of the nature of science: A historical perspective. In W. F. McComas (Ed.), *The nature of science in science education: Rationales and strategies* (331-350). Dordrecht: Kluwer.

Leisen, J. (2009). Erkenntnistheorie im Physikunterricht. *Der mathematische und naturwissenschaftliche Unterricht 62*(7), 388-394.

Lienert, G. A., & Raatz, U. (1998). *Testaufbau und Testanalyse*. Weinheim: Psychologie Verlags Union.

Mason, S. F. (1974). *Geschichte der Naturwissenschaft in der Entwicklung ihrer Denkweisen*. Stuttgart: Kroener.

Mayer, R. E. (1984). Aids to text comprehension. *Educational Psychologist 19*(1), 30.42.

Mayer, R. E. (2008). *Learning and Instruction*. Upper Saddle River, NJ: Pearson.

McClelland, D. C. (1973). Testing for competence rather than for "intelligence". *American Psychologist 28*(1), 1-14.

McComas, W. F. (2008). Seeking historical examples to illustrate key aspects of the nature of science. *Science & Education 17*(2-3), 249-263.

McComas, W. F., Clough, M. P, & Almazroa, H. (1998). The role and character of the nature of science in science education. In W. F. McComas (Ed.), *The nature of science in science education: Rationales and strategies* (3-39). Dordrecht: Kluwer.

McComas, W. F., & Olson, J. K. (1998). The nature of science in international science education standard documents. . In W. F. McComas (Ed.), *The nature of science in science education: Rationales and strategies* (41-52). Dordrecht: Kluwer.

Meyling, H. (1990). *Wissenschaftstheorie im Physikunterricht der gymnasialen Oberstufe: Das wissenschaftstheoretische Schülervorverständnis und der Versuch seiner*

Veränderung durch expliziten wissenschaftstheoretischen Unterricht (Unpublished doctoral dissertation). Universität Bremen, Bremen.

Meyling, H. (1997). How to change students' conceptions of the epistemology of science. *Science & Education 6*(4), 397-416.

Millar, R., & Wynne, B. (1988). Public understanding of science: From contents to processes. *International Journal of Science Education 10*(4), 388-398.

Mullis, I. V.S., & Martin, M. O. (1998). Item analysis and review. In M. O. Martin & D. L. Kelly (Eds.), *TIMSS Technical Report Volume II: Implementation and Analysis* (101-110). Chestnut Hill, MA: International Association for the Evaluation of Educational Achievement (IEA).

National Research Council [NRC] (1996). *National Science Education Standards*. Washington, DC: National Academy Press.

Nawrath, D. (2007). *Newtons Experimente zur prismatischen Farbzerlegung*. Retrieved from http://www.histodid.uni-oldenburg.de/26800.html

Neumann, K., Fischer, H. E., & Kauertz, A. (2010). From PISA to educational standards: The impact of large-scale assessments on science education in Germany. *International Journal of Science and Mathematics Education, 8*(3), 545-563.

Neumann, K., Kauertz, A., Lau, A., Notarp, H., & Fischer, H. E. (2007). Die Modellierung physikalischer Kompetenz und ihrer Entwicklung. *Zeitschrift für Didaktik der Naturwissenschaften 13*, 103-123.

Neville, G. (1962). The discovery of Boyle's law, 1661-62. *Journal of Chemical Education 39*(7), 356-359.

Noddack, I. (1934a). Das Periodische System der Elemente und seine Lücken. *Angewandte Chemie 47*(20), 301-305.

Noddack, I. (1934b). Über das Element 93. *Angewandte Chemie 47*(37), 653-656.

Organisation for Economic Cooperation and Development [OECD] (1999). *Measuring student knowledge and skills: A new framework for assessment*. Paris: OECD.

Organisation for Economic Cooperation and Development [OECD] (2005). *PISA 2003 technical report*. Paris: OECD.

Organisation for Economic Cooperation and Development [OECD] (2006). *Assessing scientific, reading and mathematical literacy: A framework for PISA 2006*. Paris: OECD.

Organisation for Economic Cooperation and Development [OECD] (2009). *PISA 2006 technical report*. Paris: OECD.

Osborne, J., Collins, S., Ratcliffe, M., Millar, R., & Duschl, R. (2003). What "Ideas about Science" should be taught in school science? A Delphi study of the expert community. *Journal of Research in Science Teaching 40*(7), 692-720.

Popper, K. R. (1976). *Logik der Forschung* (6th ed.). Tübingen: Mohr.

Prenzel, M., Gogolin, I., & Krüger, H.-H. (2007). Editorial. In M. Prenzel, I. Gogolin & H.-H. Kröger (Eds.), *Kompetenzdiagnostik: Zeitschrift für Erziehungswissenschaft Sonderheft 8/2007* (5-8). Wiesbaden: VS Verlag für Sozialwissenschaften.

Prenzel, M., Schöps, K., Rönnebeck, S., Senkbeil, M., Walter, O., Carstensen C. H., & Hammann, M. (2007). Naturwissenschaftliche Kompetenz im internationalen Vergleich. In M. Prenzel, C. Artelt, J. Baumert, W. Blum, M. Hammann, E. Klieme, & R. Pekrun (Eds.), *PISA 2006: Ergebnisse der dritten internationalen Vergleichsstudie* (63-105). Münster: Waxmann.

Priemer, B. (2003). Ein diagnostischer Test zu Schüleransichten über Physik und Lernen von Physik – eine deutsche Version des Tests „Views About Science Survey". *Zeitschrift für Didaktik der Naturwissenschaften 9*, 160-178.

Reble, A. (2004). *Geschichte der Pädagogik*. Stuttgart: Klett Cotta.

Reif- Acherman, S. (2004). Heike Kamerlingh Onnes: Master of experimental technique and quantitative research. *Physics in Perspective 6*(2), 197-223.

Reiss, K., Heinze, A., & Pekrun, R. (2007). Mathematische Kompetenz und ihre Entwicklung in der Grundschule. In M. Prenzel, I. Gogolin & H.-H. Kröger (Eds.), *Kompetenzdiagnostik: Zeitschrift für Erziehungswissenschaft Sonderheft 8/2007* (107-127). Wiesbaden: VS Verlag für Sozialwissenschaften.

Roberts, D. A. (2007). Scientific literacy/ Science Literacy. In S. K. Abell & N. G. Lederman (Eds.), *Handbook of research on science education* (729-780). Mahwah, NJ: Lawrence Erlbaum Associates.

Rosenberger, F. (1965). *Geschichte der Physik in Grundzügen: Geschichte der Physik in den letzten hundert Jahren*. Hildesheim: Olms.

Rost, J. (2004). *Lehrbuch Testtheorie – Testkonstruktion*. Bern: Hans Huber.

Roth, H. (1971). *Pädagogische Anthropologie* (Vol 2). Hannover: Schroedel.

Rutherford, E. (1914). On the structure of the atom. *Philosophical Magazine 27*, 488-498.

Rutherford, E. (1920). Bakerian Lecture: Nuclear Constitution of Atoms. *Proceedings of the Royal Society A (97)*, 374-400

Scharf, V. (2004). G. S. Ohms Weg der Erkenntnis – das Wechselspiel von Empirie und Theorie. In C. Hößle, D. Höttecke & E. Kircher (Eds.), *Lehren und lernen über die*

Natur der Naturwissenschaften (148-161). Baltmannsweiler: Schneider Verlag Hohengehren.

Schecker, H., Fischer, H. E., & Wiesner, H. (2004). Physikunterricht in der gymnasialen Oberstufe. In H.-E. Tenorth (Ed.), *Kerncurriculum Oberstufe II: Biologie, Chemie, Physik, Geschichte, Politik* (148-234). Weinheim: Beltz.

Schecker, H., & Parchmann, I. (2006). Modellierung naturwissenschaftlicher Kompetenz. *Zeitschrift für Didaktik der Naturwissenschaften 12*, 45-66.

Schirra, N. (1991). *Die Entwicklung des Energiebegriffs und seines Erhaltungskonzepts: Eine historische, wissenschaftstheoretische, didaktische Analyse.* Frankfurt am Main: Harri Deutsch.

Schneider, W., Schlagmüller, M., & Ennemoser, M. (2007). *Lesegeschwindigkeits- und verständnistest für die Klassenstufen 6-12 (LGVT 6-12).* Göttingen: Hogrefe.

Schmidt, M. (2008). *Kompetenzmodellierung und –diagnostik im Themengebiet Energie der Sekundarstufe I.* Berlin: Logos.

Schommer, M. (1990). Effects of beliefs about the nature of knowledge on comprehension. *Journal of Educational Psychology 82*, 498-504.

Schommer, M. (1994). Synthesizing Epistemological Belief Research: Tentative Understandings and Provocative Confusions. *Educational Psychology Review 6*(4), 293-319.

Schreier, W. (Ed.). (1984). *Biographien bedeutender Physiker: Eine Sammlung von Biographien.* Berlin: Volk und Wissen.

Schreier, W. (Ed.) (2002). *Geschichte der Physik: Ein Abriss.* Berlin: Verlag für Geschichte der Naturwissenschaften und der Technik.

Schwartz, R., Lederman, N. G., & Lederman, J. S. (2008). *An instrument to assess views of scientific inquiry: The VOSI questionnaire.* Paper presented at the annual meeting of the National Association for Research in Science Teaching, Baltimore, MD.

Sharov, A. S., & Novikov, I. D. (1994). *Edwin Hubble: Der Mann, der den Urknall entdeckte.* Basel: Birkhäuser.

Simonyi, K. (2004). *Kulturgeschichte der Physik: Von den Anfängen bis heute.* Frankfurt am Main: Harri Deutsch.

Smith, R. M., Schumacker, R. E., & M. J. Bush (1998). Using item mean squares to evaluate the Rasch model. *Journal of outcome measurement 2*(1), 66-78.

Sodian, B., Thoermer, C., Kircher, E., Grygier, P., & Günther, J. (2002). Vermittlung von Wissenschaftsverständnis in der Grundschule. In M. Prenzel & J. Doll (Eds.), *Bil-*

dungsqualität von Schule: Schulische und außerschulische Bedingungen mathematischer, naturwissenschaftlicher und überfachlicher Kompetenzen, Zeitschrift für Pädagogik/ 45. Beiheft (192-206). Weinheim: Beltz.

Solomon, J., Duveen, J., Scot, L., & McCarthy, S. (1992). Teaching about the nature of science through history: Action research in the classroom. *Journal of Research in Science Teaching 29*(4), 409-421.

Spiel, C., & Glück, J. (2008). A model-based test of competence profile and competence level in deductive reasoning. In J. Hartig, E. Klieme & D. Leutner (Eds.), *Assessment of competencies in educational contexts* (45-65). Cambridge, MA: Hogrefe.

Sekretariat der Ständigen Konferenz der Kultusminister der Länder der Bundesrepublik Deutschland [KMK] (2004a). Bildungsstandards im Fach Deutsch für den Mittleren Schulabschluss (Jahrgangsstufe 10). Neuwied: Luchterhand.

Sekretariat der Ständigen Konferenz der Kultusminister der Länder der Bundesrepublik Deutschland [KMK] (2004b). Bildungsstandards im Fach Englisch für den Mittleren Schulabschluss (Jahrgangsstufe 10). Neuwied: Luchterhand.

Sekretariat der Ständigen Konferenz der Kultusminister der Länder der Bundesrepublik Deutschland [KMK] (2004c). Bildungsstandards im Fach Mathematik für den Mittleren Schulabschluss (Jahrgangsstufe 10). Neuwied: Luchterhand.

Sekretariat der Ständigen Konferenz der Kultusminister der Länder der Bundesrepublik Deutschland [KMK] (2005a). Bildungsstandards im Fach Biologie für den Mittleren Schulabschluss (Jahrgangsstufe 10). Neuwied: Luchterhand.

Sekretariat der Ständigen Konferenz der Kultusminister der Länder der Bundesrepublik Deutschland [KMK] (2005b). Bildungsstandards im Fach Chemie für den Mittleren Schulabschluss (Jahrgangsstufe 10). Neuwied: Luchterhand.

Sekretariat der Ständigen Konferenz der Kultusminister der Länder der Bundesrepublik Deutschland [KMK] (2005c). Bildungsstandards im Fach Physik für den Mittleren Schulabschluss (Jahrgangsstufe 10). Neuwied: Luchterhand.

Tenorth, H.-E. (1994). *Alle alles zu lehren: Möglichkeiten und Perspektiven allgemeiner Bildung.* Darmstadt: Wissenschaftliche Buchgesellschaft.

Tenorth, H.-E. (2003). Bildungsziele, Bildungsstandards und Kompetenzmodelle – Kritik und Begründungsversuche. *Recht der Jugend und des Bildungswesens 2*, 156-164.

Thomas, G., & Durant, J. (1987). Why should we promote the public understanding of science? In M. Shortland (Ed.), *Scientific literacy papers* (1-14). Oxford: Department of External Studies.

Urhahne, D., Kremer, K., & Mayer, J. (2008). Welches Verständnis haben Jugendliche von der Natur der Naturwissenschaften? Entwicklung und erste Schritte zur Validierung eines Fragebogens. *Unterrichtswissenschaft 36*(1), 71-93.

U.S. Congress (1994). *Goals 2000: Educate America Act.* Retrieved from http://www.ed.gov/legislation/GOALS2000/TheAct/

Vosniadou, S. (1992). Knowledge acquisition and conceptual change. *Applied Psychology: An International Review 41*(4), 347-357.

Walpuski, M., Kampa, N., Kauertz, A., & Wellnitz, N. (2008). Evaluation der Bildungsstandards in den Naturwissenschaften. *Der mathematische und naturwissenschaftliche Unterricht 61*(6), 323-326.

Webster, C. (1965). The discovery of Boyle's law and the concept of the elasticity of air. *Archive for History of Exact Sciences 2*(6), 441-502.

Weinert, F. E. (2001a). Concept of competence: A conceptual clarification. In D. S. Rychen & L. H. Salganik, *Defining and selecting key competencies* (45-65). Seattle: Hogrefe.

Weinert, F. E. (2001b). Vergleichende Leistungsmessungen in Schulen – eine umstrittene Selbstverständlichkeit. In F. E. Weinert (Ed.), *Leistungsmessungen in Schulen* (17-31). Weinheim: Beltz.

Wilson, L. L. (1954). A study of opinions related to the nature of science and ist purpose in society. *Science education 38*(2), 159-164.

Wilson, M. (2005). *Constructing measures: An item response modeling approach.* Mahwah, NJ: Erlbaum.

Wirth, J., & Leutner, D. (2008). Self-regulated learning as a competence. *Zeitschrift für Psychologie/ Journal of Psychology 216*(2), 102-110.

Wu, M., Adams, R., & Haldane, S. (2007). ConQuest [computer software]. Melbourne: Australian Council for Educational Research.

Wu, M., Adams, R., & Wilson, M. (2007). ACER ConQuest version 2.0: Generalised item response modeling software. Camberwell: ACER Press.

Yen, W. M., & Fitzpatrick, A. R. (2006). Item response theory. In R. L. Brennan (Ed.), *Educational measurement* (111-153). Westport, CT: American Council on Education and Praeger.

Bisher erschienene Bände der Reihe *„Studien zum Physik- und Chemielernen"*

ISSN 1614-8967   (vormals *Studien zum Physiklernen* ISSN 1435-5280)

64   Jasmin Neuroth: Concept Mapping als Lernstrategie. *Eine Interventionsstudie zum Chemielernen aus Texten*
ISBN 978-3-8325-1659-8   40.50 EUR

65   Hans Gerd Hegeler-Burkhart: Zur Kommunikation von Hauptschülerinnen und Hauptschülern in einem handlungsorientierten und fächerübergreifenden Unterricht mit physikalischen und technischen Inhalten
ISBN 978-3-8325-1667-3   40.50 EUR

66   Karsten Rincke: Sprachentwicklung und Fachlernen im Mechanikunterricht. *Sprache und Kommunikation bei der Einführung in den Kraftbegriff*
ISBN 978-3-8325-1699-4   40.50 EUR

67   Nina Strehle: Das Ion im Chemieunterricht. *Alternative Schülervorstellungen und curriculare Konsequenzen*
ISBN 978-3-8325-1710-6   40.50 EUR

68   Martin Hopf: Problemorientierte Schülerexperimente
ISBN 978-3-8325-1711-3   40.50 EUR

69   Anne Beerenwinkel: Fostering conceptual change in chemistry classes using expository texts
ISBN 978-3-8325-1721-2   40.50 EUR

70   Roland Berger: Das Gruppenpuzzle im Physikunterricht der Sekundarstufe II. *Eine empirische Untersuchung auf der Grundlage der Selbstbestimmungstheorie der Motivation*
ISBN 978-3-8325-1732-8   40.50 EUR

71   Giuseppe Colicchia: Physikunterricht im Kontext von Medizin und Biologie. *Entwicklung und Erprobung von Unterrichtseinheiten*
ISBN 978-3-8325-1746-5   40.50 EUR

72   Sandra Winheller: Geschlechtsspezifische Auswirkungen der Lehrer-Schüler-Interaktion im Chemieanfangsunterricht
ISBN 978-3-8325-1757-1   40.50 EUR

73   Isabel Wahser: Training von naturwissenschaftlichen Arbeitsweisen zur Unterstützung experimenteller Kleingruppenarbeit im Fach Chemie
ISBN 978-3-8325-1815-8   40.50 EUR

74   Claus Brell: Lernmedien und Lernerfolg - reale und virtuelle Materialien im Physikunterricht. *Empirische Untersuchungen in achten Klassen an Gymnasien (Laborstudie) zum Computereinsatz mit Simulation und IBE*
ISBN 978-3-8325-1829-5   40.50 EUR

75   Rainer Wackermann: Überprüfung der Wirksamkeit eines Basismodell-Trainings für Physiklehrer
ISBN 978-3-8325-1882-0   40.50 EUR

76   Oliver Tepner: Effektivität von Aufgaben im Chemieunterricht der Sekundarstufe I
ISBN 978-3-8325-1919-3   40.50 EUR

Alle erschienenen Bücher können unter der angegebenen ISBN direkt online (http://www.logos-verlag.de) oder per Fax (030 - 42 85 10 92) beim Logos Verlag Berlin bestellt werden.